# *CLOCKS AND CLOCK REPAIRING*

**2ND EDITION**

# *CLOCKS AND CLOCK REPAIRING*

## 2ND EDITION

ERIC SMITH

SECOND EDITION
FIRST PRINTING

First published 1979 by Lutterworth Press
Revised and enlarged edition published 1988 by Lutterworth Press

Printed in the United States of America



Library of Congress Cataloging-in-Publication Data

Smith, Eric, 1940-
Clocks and clock repairing / by Eric Smith. — 2nd ed.
p. cm.
Rev. ed. of: Clocks & clock repairing. c1979.
Includes bibliographical references.
ISBN 0-8306-0276-3 — ISBN 0-8306-0235-6 (pbk.)
1. Clocks and watches—Repairing and adjusting. I. Smith,Eric, 1940- Clocks & clock repairing. II. Title.
TS547.S58 1989
681.1'13'028-dc20 89-20341
CIP

TAB BOOKS Inc. offers software for sale. For information and a catalog, please contact TAB Software Department, Blue Ridge Summit, PA 17294-0850.

Questions regarding the content of this book should be addressed to:

Reader Inquiry Branch
TAB BOOKS Inc.
Blue Ridge Summit, PA 17294-0214

# CONTENTS

# LIST OF PLATES

*(Plates 1–16 between pages 60 and 61)*

## LIST OF DIAGRAMS

# INTRODUCTION

It is almost proverbial—particularly between parents and children—that you should not take complex things apart in case you cannot put them together again, and this axiom is applied especially to clocks. The assumption is that it will cost more money to undo your misadventures than it would have cost to have had the original fault properly corrected, and that it will be embarrassing or humiliating to present the *disjecta membra* to the professional. Both assumptions may well be correct. Some people will always ignore such warnings, however, and this book is for those who feel so inclined, in the hope that it may make the way a little safer.

For whatever reason, you may find yourself intrigued by the idea of working on clocks or a clock. You can proceed by hit and miss methods, but you may well do more damage than you cure and you will prevent few of the faults which have yet to appear. Before working on a clock you really need to know what type of clock it is, what general principles it follows, where it differs from the general types and why, how many of them and in what order you will have to remove the various parts. Often, for instance, you cannot correct a fault in the striking side without knowing something of the 'going' side, and vice versa. You cannot correct for worn parts if you do not know what their exact shape should be, or in which direction the wheels turn. It will serve little purpose repairing or replacing an escapement if the mainspring or train is wasting power before the escapement is reached. The impulse to oil everything because of a certain sluggishness may be strong but, as a sole cure, it is generally wrong and can do harm. And so on. This sort of general information will be found, after a few remarks about tools and materials, in the first five chapters of this book, along with diagrams of the parts and movements which you may come across.

There are those who do not regard electric clocks as in the province of the clock repairer. A considerable proportion of the clocks in existence must, by now, be electrically operated. and it is a proportion which will assuredly increase. If you do not wish to be concerned with this fast-developing area of horology, that is your affair, and it is true that many modern electric clocks do not lend themselves to extensive repair. There will be others, however, and not only those with an interest in electricity or electronics, who will regard electric clocks as being as much their concern as weight- and spring-driven clocks. Therefore a survey of some types of electric clock is given in Chapter 5.

The rest of the book is taken up with practical matters of cleaning, diagnosing faults, and repairs which you can make. There is much in horology, as in any craft, which cannot be taught, let alone self-taught from a book, but books can stimulate interest and be useful for reference, and it is hoped that this book may partly serve these purposes.

I feel bound to notice two changes among many which have taken place in the world of horology since the first edition was published. First, the 'quartz revolution', which may prove almost as significant as the introduction of the pendulum three hundred years ago, has occurred. Except for occasional excellence, the mechanical clock has been largely superseded for domestic time-keeping, replaced by types of clocks whose economics of repair are very different. Moreover mechanical clocks are increasingly bought for aesthetic and reasons other than for time-keeping. Secondly amateur interest has grown greatly and both parts and special services (such as dial restoration) are available far more widely than would have been conceivable even ten years ago. Both of these changes are having a marked influence on the hobby and, indeed, its boundaries with the trade. More people take their interest further, and it is partly for this reason that a new chapter, introducing more advanced work, has been added to this edition.

Finally, I would like to acknowledge a debt to writers on horology, too many to name individually, and to all those whose clocks I have repaired or borrowed for inspection. I am particularly indebted to Mrs. A. Allnutt for the photographs.

*Chapter One*

# TOOLS AND MATERIALS

The list of possible tools for working on clocks is virtually endless, and many items may be used constantly. Some are, however, for the specialist who repeatedly performs a task occurring only once or twice in the careers of amateur repairers, and others are best left until they are necessary for doing a particular job to a very high standard. Much can be done with a few screwdrivers and a pair of pliers, but generally the indispensable tools include tweezers, nippers, a few punches, files, broaches and drills. In addition, a soldering iron, soft solder, a gas blowlamp and brazing solder may be useful, although the true clock-repairer prefers riveted and driven fits to soldered joints, which tend to be imprecise as well as unsightly. A light hammer is another necessity, and a small vice will come in useful.

According to the size of the clock concerned, various screwdrivers are required. It is best to have a long one, say nine inches, with a blade of about $\frac{3}{8}$ of an inch (9·5 mm), for reaching into confined cases and dealing with the larger screws, and a set of watch screwdrivers, which are readily available, for smaller jobs.

Small pointed pliers are most used, and a larger pair is already in most household tool kits. Avoid the electrician's fine pliers with a very long nose, for they are inclined to be too springy and to go out of alignment. Nippers are used for cutting wire, particularly the brass or steel pins used for holding most clock movements together. The top-cutting type, in a small size, are most useful. Large and small tweezers with pointed ends are used for picking up small, particularly cleaned, objects and for positioning them. It is pointless to economize with pliers, tweezers or nippers—far better to have a really good pair of each, with firm joints and accurately aligned blades, than to buy cheaper types which will slip and spoil whatever they are holding or snap the part out and flick it across the room. Brass tweezers are especially useful since they avoid the trouble of magnetism; non-magnetic nickel steel tweezers are also available.

Punches are used for dislodging stubborn pins, for spreading metal, and also in riveting and bushing. Simple punches with flat or rounded ends are easily made from tool steel, and the variety of sizes likely to be needed over the years suggests the policy of making them for each job as required. Hollow punches for riveting over wheel arbors are best bought in an assortment if needed. Sets of elaborately shaped punches for many tasks can be purchased, but are fairly expensive and generally more applicable to work on watches. Hole-closing punches (being hollow with sharp edges and a sprung central locating pin) are useful for quick work on cheap movements but their use elsewhere is frowned upon, for good reasons, and they are not required for good quality work.

Files and broaches are the basic cutting implements, save for a fine hacksaw which will be useful if you contemplate cutting a new part. Files can be bought in assortments and are needed in several grades, shapes and sizes. It is as well to have at least one fine flat file with smooth edges (to avoid cutting adjacent metal), two or three round needle files, and a square or triangular file. Half round files are very useful for work on wheel crossings (spokes). New files are best for brass, and old worn files can be used on steel. Broaches are tapered lengths of five-sided bright steel used for enlarging holes. They are sold in sets, often with tap-wrenches or chuck-handles to rotate them in the work, and it will be best to buy according to the general size of holes in a particular movement, for those used in a fine French movement are of little use in the holes of a heavy grandfather (long-case) movement.

A basic set of fine twist drills in a hand drill will do much for larger movements. For finer work, fluted or spear-headed clock drills are necessary, either in an Archimedian drill-stock or in a small electric powered unit. Drills are becoming expensive, and certainly a powered unit and flexible drive will cost more than a single repair job would justify, though drills can in fact be made. The truth is, however, that in the great run of humdrum repairs a drill is not required, it being far more often the case that one has rather to open an existing hole with broach or file than to make a new one, and, as most blind holes are screw-holes, in a clock they can be enlarged with a twist drill. Portable electric drills are not suited to work on a clock movement, except perhaps as power units in cleaning larger movements, for they are far too heavy and have excessive play in their bearings.

Enlarging an existing blind hole is often occasioned by the need to fit a new and slightly larger screw, and here of course taps and a tap wrench are required. These are costly, and the repairer will have to decide whether to go for an assortment (which may be useful in other jobs around the house) or to buy one or two taps for a particular job and known replacement screw. It should be noted also that threads on many old movements do not exactly coincide with modern gauges and it may be preferable to use a modern, but not unsightly, screw and newly tapped hole to replace a missing screw or stripped thread.

The list of materials again could be very long, but for the common repair it can be substantially reduced. Pegwood is cheaply bought by the bundle. It is used for cleaning pivot holes—for which purpose matchsticks are too soft. Brass bushes are needed for repairing worn holes, and a mixed packet of appropriate size will have to be obtained. Tapered clock pins, usually of steel, can be bought by the gross in mixed sizes; old pins can be used to a certain extent and new ones can also be filed from wire, but it takes time. A collection of screws is a useful standby and can be assembled over the years, but packets of watch, alarm, carriage clock screws (and so on) are available. Clock or watch oil is essential, and cheap in the very small quantities used—a small bottle will last years if kept well stoppered and out of bright light. It, rather than the lightest machine oil, must be employed, since the latter quickly thickens on slow-moving parts; even then, clock oil must not be used on gear teeth, being reserved for pivots and other moving surfaces. For movements with a bright finish, metal polish can be used once gross dirt and tarnish have been removed. The particular brand is of little importance and everyone has his own preference. The dully finished plates of many antique clocks, particularly English ones, are best scrubbed with tripoli and oil—tripoli is available very cheaply in a block—which will give a purer but less glazed finish than will metal polish. The only satisfactory material for cleaning pinions is oilstonedust, which is difficult to come by: fine emery paper or powder are reasonable substitutes. For basic cleaning, household ammonia and either soapflakes or washing-up liquid are as effective as many of the proprietary solutions on the market. For cleaning cases, the materials must depend on the work in hand. Turpentine (useful also as a lubricant when drilling metal) and beeswax may be used on waxed wooden and imitation marble or slate cases. There are also special preparations for marble cases. Button polish is convenient for touching up french polished or varnished woodwork. In general, a switch-cleaner is a handy solvent for cleaning metal and synthetic surfaces but has to be used carefully in case it dissolves the material. It is not usually safe to use on dials since it is likely to make the markings unstable. Clear lacquer is needed for cleaned external brass parts and can be had in a bottle for use with cottonwool, or in an aerosol, which gives a dappled, less clear finish. Liquid gold leaf, or a similar preparation, is worth having on the shelf for touching up ornaments, but sizable gilded parts and cases have to be professionally gilded as a rule. 'Treasure gold' wax (from most art shops) is satisfactory on less valuable gilt pieces. 'Craftsman' and 'Connoiseur' are similar products. There is of course considerable scope for ingenuity in the choice of further household materials for cleaning movements and cases.

Here are listed most of the tools and materials which you will collect if you think that rejuvenating your first clock may be but the start of a habit or a pastime. Nonetheless, attending to a clock is often a matter of degree. One may wish to do no more than to 'get it going again', or to restore it virtually to new condition and appearance. It depends on the time and the money available, the value of the clock and for whom you are doing the work. We are concentrating in this book more on the immediate repair than on the full restoration, although the immediate repair includes cleaning and general overhaul. It is almost certain that with very meagre equipment – screwdrivers, fine files, oil, pliers – you can persuade your clock to keep reasonable time and to sound correctly. To ensure that it stays in this condition for long and without attention entails more extensive work, and correspondingly more will be needed in the way of tools and materials. We consider finally more expensive tools in which you may come to think of investing.

Sometimes difficulty is encountered by the newcomer to horology in obtaining tools, materials and replacement parts. It is true that some suppliers will deal only with the trade, often in part because it is not economic for them to handle small and irregular orders. However, this problem is not as great as it was a few years back.

The clock materials supply business may appear in the Yellow Pages telephone directory under the headings 'Clock and Watch Importers and Wholesalers' and 'Jewellers Supplies' or 'Clock and Watch Spare Parts'. Reference to the journal (*Antiquarian Horology*) of the Antiquarian Horological Society (High Street, Ticehurst, Wadhurst, Sussex) will occasionally provide some help on materials, and more particularly on specialist work (e.g. restoring dials), as will the *Horological Journal* (Brant Wright Associates Ltd., PO Box 22, Ashford, Kent) and *Clocks* (Model and Allied Publications, Ltd.). A selective list of suppliers is included on p. 192.

Finally, whilst for many reasons its importance as a distribution and manufacturing centre for clocks and watches is much reduced, Clerkenwell in London still contains a number of tool and material suppliers to the clock trade; if you walk around Hatton Garden, Clerkenwell Road and St John Street, you will be able to buy for cash all that you need in the way of tools and materials—and you will also find much that you do not need but whose purchase it is hard to resist.

*Chapter Two*

# HOW CLOCKS WORK

When you have some basic equipment, you need to consider what is the rationale behind the apparent jumble of wheels and levers which confronts you when you remove the hands and dial of your clock, or get access behind them by taking it out of the case. In pondering this, you arrive at some of the principles of clock workings which it is necessary to understand if a clock is to be properly adjusted.

### *Lay-out of Gear Trains*

Perhaps the first thing is the concept of gearing. The minute hand of a clock revolves once in an hour and the hour hand revolves, as a rule, once in twelve hours, that is, twelve times as slowly. These two hands are connected by wheels, known as motion wheels, which are usually directly behind the dial. Now the basis of gearing is that if two rollers revolve against each other, their relative speeds of revolution will be proportionate to their sizes—if one is half the size of a bigger roller, the bigger one will revolve at half the speed of the smaller one, or, from the other point of view, the smaller one will revolve at twice the speed of the bigger. Gear wheels are rollers with teeth so that they do not slip and, as the teeth on each roller or wheel are the same size (so that they mesh properly), it follows that we can express the size of the wheels by the numbers of teeth on them as readily as by the size of the wheels which they roughly represent. Thus our 1:2 roller ratio may be represented as a pinion of 12 teeth and a wheel of 24 teeth.

In the case of the clock hands we could have a minute hand on the arbor having a pinion with twelve teeth and an hour hand on the arbor of a wheel with 144 teeth (12 times as many, revolving at 1/12 the speed). However, two rollers revolving against each other, one driving the other, must revolve in opposite directions, yet we know that both clock hands revolve 'clockwise'. This is arranged by using an intermediate wheel (known as the minute wheel), usually offset from the centre on a stud, which may or may not take part in the actual reduction gearing, but which ensures that both hands revolve in the same direction. On old clocks this wheel is the same size as the central minute wheel (the 'cannon pinion') and lets off the striking; its pinion then rotates 1–12 to a large hour wheel.

The clock movement is primarily a line of such wheels, the gear train, each wheel turning the next pinion. Normally wheels drive pinions, rather than the reverse. So far as the time (going) train is concerned the wheels carrying the hands are not at either end of the train, but in the middle. At one end is the power supply, usually a weight or spring turning a wheel or barrel, and at the other is the escapement. The escapement is there to interrupt the motion of the gear train in a precise and regular manner, so that the clock may measure out the passage of time correctly. The breaks in action are of course clear with a seconds hand, and can be seen on the minute hand if it is studied closely. The minute hand is attached to a slowly turning arbor (the centre arbor)—it cannot be attached to the slowest of all, the barrel, since that only revolves some five times during the full unwinding of the spring in a day or in a week. (In an eight day clock, an extra wheel is used to extend the duration.) The layout of the thirty hour (one day) and eight day trains is shown in Fig. 1 in diagrammatic form.

There may also be a striking train. This is quite separate from the going train, being held in check until released by a lever (the lifting piece) in touch with the centre arbor at every hour and half hour. The chiming train is similar, being let off every quarter of an hour (as a rule) by the lifting piece, and itself letting off the striking train when it has finished chiming the fourth quarter. (There are, however, also arrangements for striking quarters from the striking work without using a separate chiming train—*see* Chapter 4.) The speed of striking or chiming is controlled not by an escapement but by the air resistance of a revolving fly or fan. Nonetheless, although the striking and chiming trains do not in themselves measure time, their gear ratios and layout have still to be carefully calculated. The reason for this lies in the necessity for an exact and regular programme of strokes to be struck or chimed within a

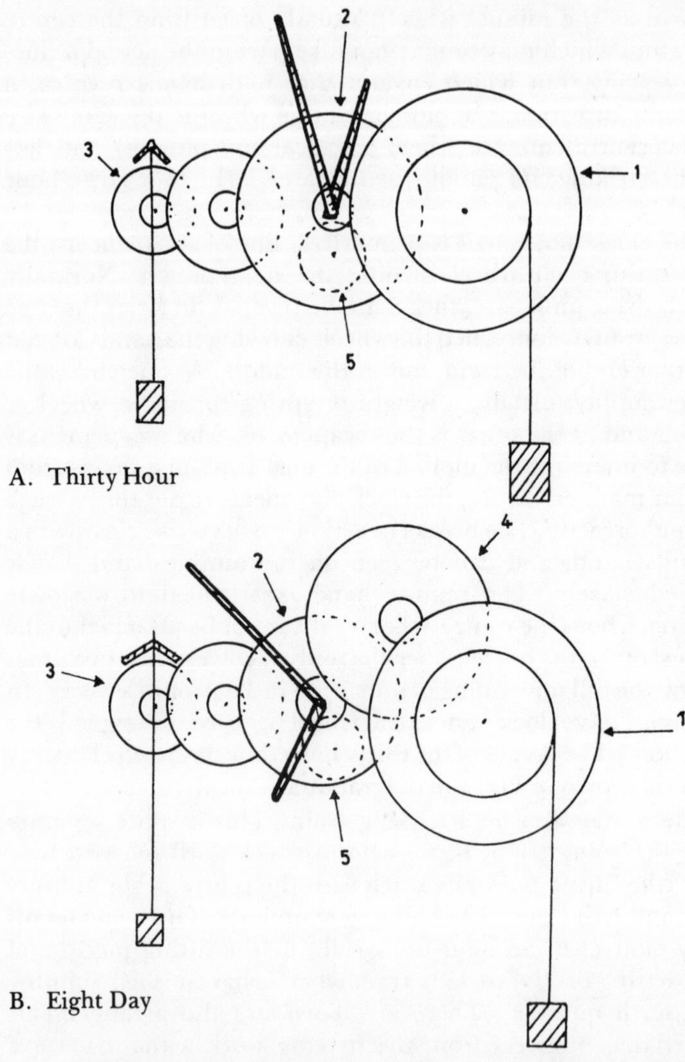


1. Power supply – barrel for weight or mainspring, or fusee
2. Centre wheel and arbor, connected to minute hand and hour hand via motion wheels
3. Scapewheel and escapement
4. Intermediate wheel (Usually absent from weight-driven clocks)
5. Motion (minute) wheel

Fig. 1. Main wheels in the going train

period. Thus, in every twelve hours, the gong or bell of an hour-striking clock must be struck exactly 78 times (1 + 2 + 3 . . . 10 + 11 + 12). The striking train can be locked only by obstructing a wheel (the locking wheel), and this wheel must revolve exactly once for each blow of the hammer, or else it would be impossible to increase the number of strokes by one at each hour. The same applies at quarters in chiming, where the chime sequences accumulate until four are struck on the hour, where the locking wheel must revolve once for each single sequence or group of notes. This relating of the quarter sequence to the revolution of the locking wheel is carried out by removable wheels, known as ratio wheels, which are usually placed outside, on the backplate in modern clocks.

It follows, of course, that though the sounding trains do not keep time, their duration of running can be precisely calculated so that the spring or weight runs down at the same time as that of the going train. The ratio of the fly to its preceding wheel (the warning wheel) does not have to be calculated, however, since this merely governs how quickly or slowly the sounding train runs whenever it is released, and there is no necessity here for the same teeth of wheel and pinion to engage at each revolution, or for the fly always to stop in the same position. In the case of an alarm train there is no complication—it is merely a matter of how long the alarm is to sound before (if uninterrupted by the wakened sleeper) its spring runs down. Diagrammatically; the layout of striking and chiming trains is shown in Fig. 2. (The function of the named wheels is explained in Chapter 4.)

The trains are of course driven from the slow-moving end by a weight or spring. This takes a relatively long time to unwind, and the smaller wheels further down the train revolve relatively fast. Nonetheless, in any ordinary mechanical clock the timekeeping is carried out by the escapement and a vibrating pendulum or balance wheel, not by the gear train itself and the load which it imposes on spring or weight. It is essential to recognize that these controllers are driven by the spring or weight and in no way drive the train, even though they govern how fast it moves. There are, however, clocks in which the pendulum or balance drives the train, and then the wheels bearing the hands are the slowest and the last in the train.

Most electric clocks are of this type. In these clocks the pendulum or balance is kept in motion by an electro-magnet switched regularly on and off by contacts or an electronic circuit. Then, a final instance of power applied to the 'fast' end of the train, there is the popular electric mains clock in which the first, fast rotating wheel is in fact the rotor of a synchronous electric motor whose speed is entirely controlled by the frequency of the AC mains supply.

Normally there is a separate spring or weight for each

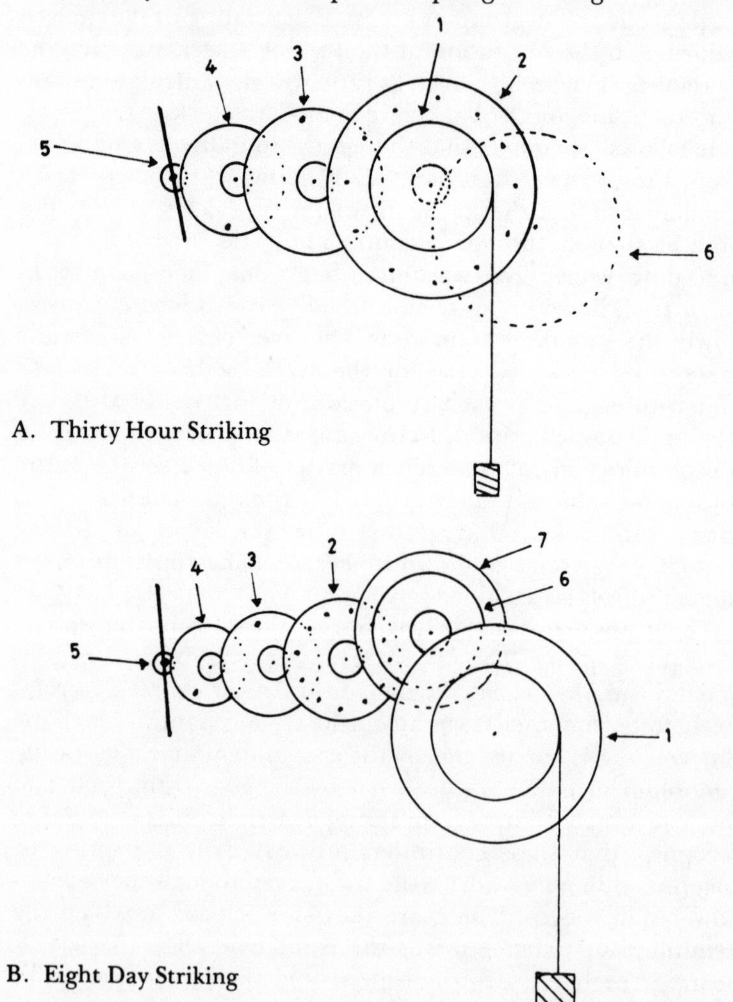

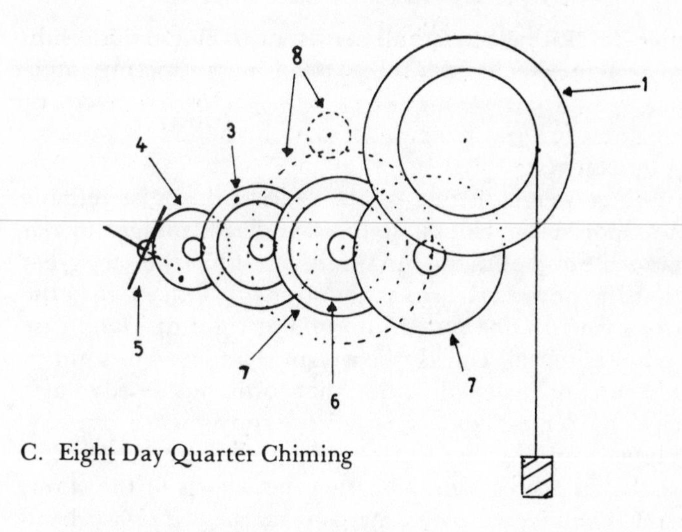


C. Eight Day Quarter Chiming

1. Power supply
2. Pin wheel or hammer wheel
3. Locking wheel (hoop wheel in the countwheel system)
4. Warning wheel
5. Fly or fan
6. Countwheel (if countwheel system)
7. Intermediate wheel(s)
8. Ratio wheels on intermediate wheel arbor and on chime barrel arbor (outside movement plates)

Fig. 2. Main wheels in typical striking and chiming trains

function—for example, going, striking and alarm trains. Mention has, however, been made of quarter-striking from the striking train and there are in fact several other common arrangements directed towards saving space and economizing in winding. The most general is the use of a single mainspring for alarm and going in modern alarm clocks and, in the electric parallel, using the synchronous motor's field magnet to operate a buzzer alarm. Striking and chiming trains have also been operated in tandem off one large mainspring, and electric striking clocks, which are not very common, may run both going and striking from the same motor. It is also possible to mix electric and mechanical power, using a regularly switched motor to wind a small mainspring, as is very common in a range of cheap electric wall clocks, or to keep wound subsidiary springs which operate striking and chiming mechanisms.

*Driving by weights*

The falling weight is one of the oldest and most reliable sources of power for clocks. It has the disadvantages of not being portable and of taking up space, but it has the very great merit that the power released is constant from the start to the finish of its run. In one system, normal to eight day long-case and good quality wall-clocks, the weight is attached to a gut or wire line which is fastened with a knot to the inside edge of a cylindrical barrel, which is grooved for convenience (Fig. 3). The weight is wound by turning the square barrel arbor. In order that this process does not turn the wheels of the train, the barrel is connected by a pawl and ratchet (click-work) to the first gear wheel, on its arbor, allowing it to turn freely in the direction of winding but driving the train when released. There may also be a subsidiary ratchet attached to an internal spring. This is kept charged during running by a long finger or detent when the clock is going, and it continues to drive the clock during winding (which would otherwise momentarily reverse the train and can damage the escapement and lead to slow time-keeping). This 'maintaining power' is only found on movements of the best quality. The same arrangement (but without maintaining power) is repeated for the striking and, if present, the chiming trains, which use progressively heavier weights.

The alternative system is and was used almost entirely on thirty-hour clocks, for instance lantern clocks, many simpler long-case clocks, some cheaper weight-driven wall clocks, including many cuckoo clocks. It employs a lighter weight (some 8 lb as against perhaps 11 lb for the going of eight day long-case clocks) and one weight only for the going and striking trains, which are nonetheless quite separate. The weight hangs from a loop in an endless chain whose links drop over spikes on the pulleys of the first wheels in the trains (Fig. 3). One of these pulleys is fixed. The other, that of the going train, is connected by a crude form of click, which engages with the spokes (crossings) of the wheel during

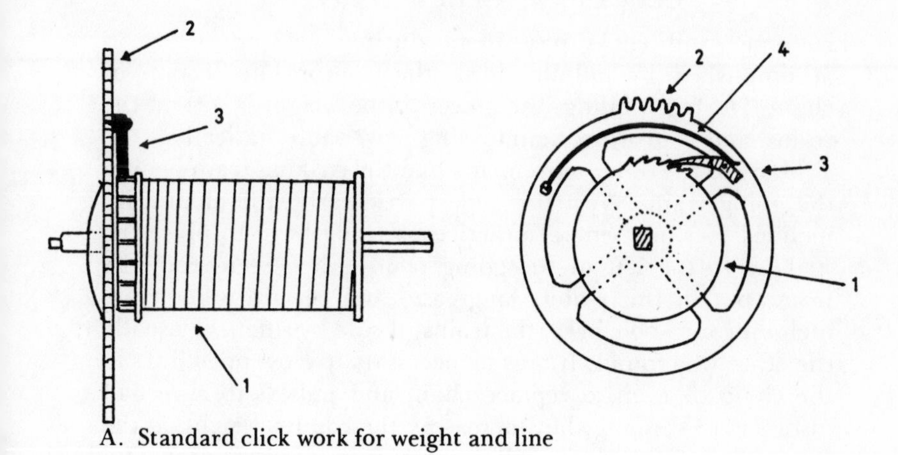


A. Standard click work for weight and line

1. Barrel attached to barrel arbor with winding square and click wheel
2. Greatwheel, free on barrel arbor, held by sprung dome washer
3. Click
4. Click spring

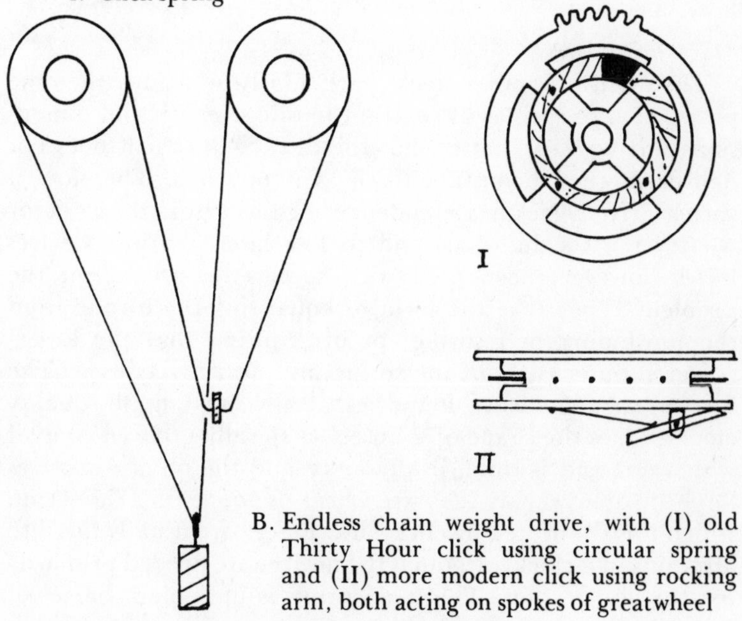


B. Endless chain weight drive, with (I) old Thirty Hour click using circular spring and (II) more modern click using rocking arm, both acting on spokes of greatwheel

Fig. 3. Weight drives and clickwork

running, and clicks past them during winding, which is accomplished by pulling down the appropriate section of chain. During winding, the power of the weight is still on the trains, and so provides maintaining power automatically.

Many of the older movements had the striking train behind the going train. With this type particularly, the chains are inclined to slip. General practice has long been to place the striking to the left of the going train, when seen from the front, but in thirty-hour long-case, and certain other types including cuckoo clocks, the trains, if side by side, are usually the other way round. It may be necessary to close open links in the chain or even to replace chain and pulleys to cure bad instances of slipping, but increasing the counterweight on the slack loop of the chain will often help towards a cure. Old rope-driven clocks use exactly the same system, with sharper spikes on the pulleys, but the ropes wear and are also inclined to slip. New ropes are not hard to splice and can be obtained, though it is thus sometimes worth converting these clocks to chain drive (*see* Chapter 7).

*Mainspring Drive*

The mainspring of a clock, particularly at a time when the production of springs of constant tensile strength and dimensions was no easy matter, has the drawback that it does not deliver power of constant force as it unwinds. Therefore a spring-driven clock has a tendency to gain early in the week (or whatever is its duration) and to lose later. Various devices have, however, been used with success to circumvent the problem. They take the form of equalizing the torque from the unwinding mainspring, or of ensuring that the lower-powered outer turns of the spring are not actively used. The former arrangement, found especially in English quality clocks, takes the shape of a 'fusee', a spirally grooved cone of which one end is small in diameter and the other almost as large in diameter as the first wheel in the train (Fig. 4), to which it is connected by clickwork. Fusees are usually fitted to all trains in a clock, although the devices are needed primarily for the going train. The mainspring is in a plain barrel of constant diameter, round which a gut or wire line, or a chain, is wound when the spring is run down. The other end of this line

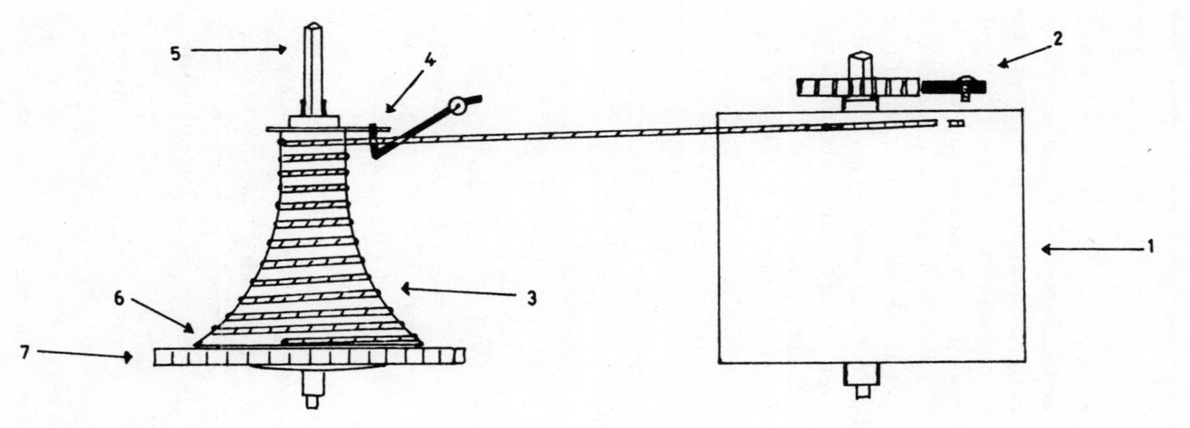


1. Mainspring barrel
2. Barrel click, ratchet and square to set up mainspring
3. Fusee with line wound on from barrel
4. Fussee stop and stop finger
5. Winding square
6. Click work (in base of fusee)
7. Greatwheel

Fig. 4. Fusee and barrel fully wound
(Note—part of a turn of line is left on the barrel to prevent the hook from being pulled out when the fusee is fully wound.)

is fixed to the large diameter base of the fusee.

The clock is wound by turning the fusee square, thus winding the line on to the fusee against the resistance of the coiled mainspring. The taut line presses a lever (the fusee iron) into the path of a finger (the fusee stop) on top of the fusee so that the mainspring cannot be overwound—otherwise the line would come off the fusee and be likely to break, and the unleashing of such a powerful mainspring would wreak great damage. As the clock runs, the spring pulls the line from the fusee, causing the latter and the train to turn, but when the spring is fully wound the line is at the narrow end of the fusee and the spring is at a mechanical disadvantage, whereas the line gradually comes to the lower part of the fusee and the lesser strength of the spring at this stage is compensated for by the effect of leverage given by the larger end of the fusee. Often maintaining power is incorporated, and it takes the form of an auxiliary spring and an additional click and clickwheel in the base of the fusee, held at tension by a finger or detent pivoted between the plates to act on the maintaining clickwheel.

The fusee is an extremely effective compensator, but it is now rarely used in domestic clocks since it is heavy, occupies space, involves extra material and manufacturing time, and there is the risk of damage from a broken line. The practical alternative is the 'going barrel', a cylindrical barrel containing the main spring and arbor, with wheel teeth at one end so that it acts as the first wheel in the train. The inner end of the spring is hooked on to the arbor, and the outer end to a hook in the wall of the barrel, the spring being completely enclosed. The spring has to be wound into its barrel by hand or with a tool called a mainspring winder. It will only go into the barrel when considerably wound, the effect being that the weak last turns are not directly used. This effect may be increased, though not generally in English clocks, by a device known as stopwork, of which starwheel stopwork is the commonest form, but of which there are various types involving irregular wheels (Fig. 5). Here a projection fixed to the barrel arbor probes the irregular teeth of a small wheel on the barrel face, advancing it one tooth at each revolution, but unable to pass at the one raised tooth. This means both that the spring cannot be overwound, and also that it cannot run down beyond a point

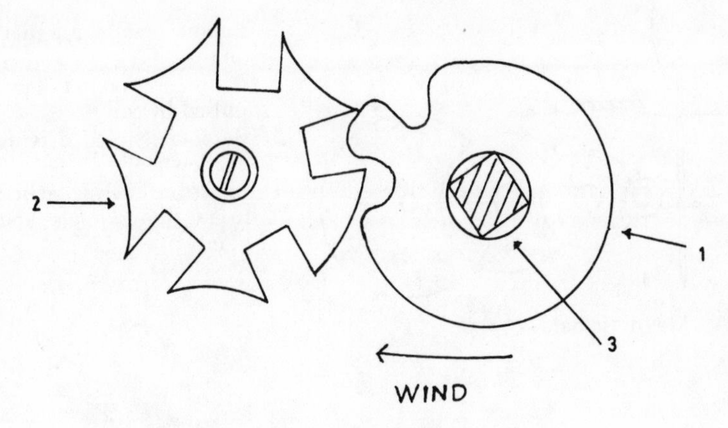


1. Stop finger fixed to barrel arbor
2. Irregular star wheel screwed freely to barrel
3. Winding square

Fig. 5. Star wheel stopwork (shown fully wound)

determined when the clock is assembled, for of course the projection will baulk at the eccentric part of the wheel whichever way it is revolving. In practice the spring is set up by about one turn before the stopwheel is fitted to the barrel, so that this revolution (together with those rendered ineffective by the constraint of the barrel) is not used to turn the wheels of the train.

Springs in barrels should not occupy more than half the area of the barrel; if it takes up more, a spring cannot be fully wound or, if much less, the weakest part of the spring will be used. The strength of the spring depends on the height and, more particularly the thickness, of the metal used.

It is usual for a clickwheel with a square hole to be fitted over the going barrel so that it engages with a click and spring mounted on the front or back plate of the movement. Thus, in winding, the arbor is turned and the clickwork operates whilst, in running, the barrel revolves and the arbor and clickwork stay fixed. The square arbor of the mainspring barrel of a fusee system is, however, never turned, nor is its click moved, once the fusee has been set up on assembly of the clock, for the clickwork is part of the fusee and it is the fusee arbor which is

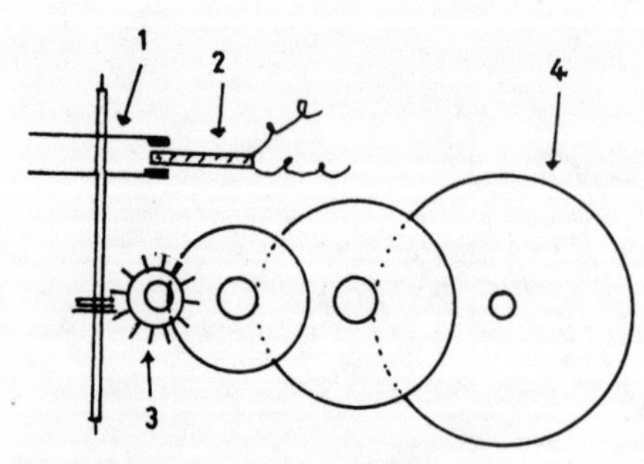


1. Balance with magnets on rims
2. Balance magnets impulsed by coil
3. Balance staff driving 'scapewheel'
4. Centre wheel attached to motion wheels and hands

A. Magnetic balance type

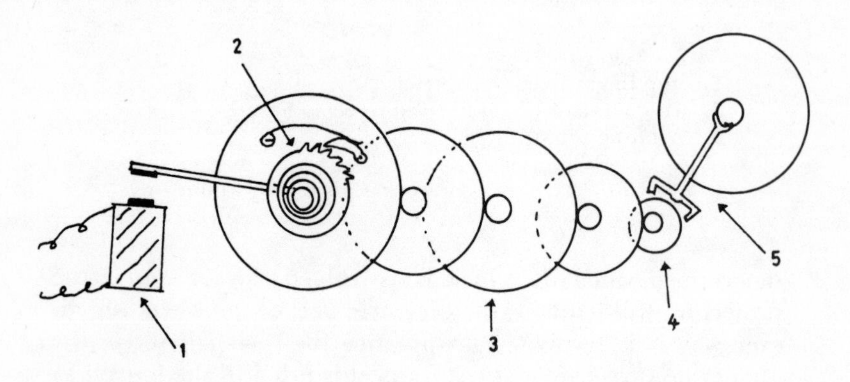


B. Re-wind type

1. Electro-magnet energized by contacts on great wheel
2. Mainspring wound by electro-magnet
3. Centre wheel attached to motion wheel and hands
4. Scapewheel
5. Balance and escapement

Fig. 6. Lay-out of modern battery movements

used for winding. Often, especially in cheaper movements, an open spring is used without a barrel. The first wheel is loose on the main spring arbor and has a click riveted to it, with a click-spring, in such a way that it presses on the clickwheel fixed to the arbor. Thus again, the arbor is turned to wind, and the wheel turns only during running. The inner end of the spring is hooked to its arbor, and the outer end is secured to a movement pillar or special stud for the purpose. The positions of these fixtures determine how far the spring can unwind. Usually, there is spare power in the spring used so that it does not fully unwind in the nominal duration of the twenty-four hours or seven days going of most ordinary clocks.

*Electricity*

The special arrangements of (*see* Fig. 6) electric clocks are discussed in Chapter 5.

*Functions of the Escapement*

Looking at the plan of the movement as a whole, we have a situation in which (save in most electric clocks) power is applied to one end of a train at winding and is gradually released at the other end. Since, although it does not drive the clock, it controls the speed of release, and hence the time shown, the escapement is at the heart of the mechanical clock.

There are many forms of escapement, some of which will be examined later in Chapters 3 and 7, but all are essentially interrupting devices and they are governed by a pendulum or a balance wheel and balance spring. Both of these oscillators have a natural frequency of vibration, that of the pendulum being determined by its length, and that of the balance by a complex of factors including the size and mass of the wheel and its balance spring. Both are theoretically, though in practice not entirely, isochronous—that is, no matter how large the swing of the pendulum or vibration of the balance wheel, any one oscillation will always take place in the same time. Thus, to a limited extent, they are independent of variations in the power applied to them—although, quite apart from the other qualities of a particular oscillator, this independence is limited by the type of escapement used. They will also keep vibrating with very little power, provided that an impulse is given regularly and at the correct point in their vibration.

The escapement not only holds and releases the train according to the vibrations of pendulum or balance, but also at each vibration imparts the small impulse necessary to keep them in vibration. If insufficient power reaches the oscillator because of a defect in the train or the mainspring, or because power is wasted in an ill-adjusted escapement, the pendulum or balance, and the clock, will eventually stop. If, for similar reasons, there is too much power the clock will keep bad time and also will rapidly wear out.

*Turning Clock Hands*

Finally, we return to the hands and face behind which this mechanism beats away. Obviously, if your clock stops, it is possible to wait twelve or twenty-four hours and then to start it again, but this is inconvenient. Therefore a frictional set-hands device, a sort of clutch, is built in so that the hands can be set to time without forcing the train to revolve with them, which could only result in broken hands or a broken escapement, or both. Various arrangements have been used (Fig. 7). They depend on the frictional connection of the minute hand to its arbor. As the hour hand is itself geared to the minute hand by the motion wheels, as we have seen, it makes an appropriate movement when the minute hand is advanced. There is much scope here for adjustment. In some arrangements too tight fixture of the minute hand will stop the clock—the pin itself should fit tightly, but thickness of the washer (collet) behind can be varied. Too loose a tensioning spring will not stop the clock, but it will result in vagaries in its indication of time—though the movement may itself be keeping good time—and it may lead to failure to strike.

The word 'advanced' is used deliberately of setting hands in the sense of moving forward. It is something of an old wives' tale that a clock hand must never be turned backwards, for there are types of clock where little harm will result, though if the clutch is tight the reversal of the train can cause damage. But in general it remains a good rule. There is always risk, and in most striking or chiming clocks there is great danger, in pushing hands backwards. This arises because of the lifting piece which sets off the strike. It is operated by pins or cams on the centre (cannon) pinion or minute wheel and when the strike sounds, the lifting piece falls off a pin as the hands turn past the hour. If the hand is pushed backwards, the pin may then be forced into the lifting piece, so bending the piece, possibly breaking the pin and ruining the striking. Flexible lifting pieces are sometimes employed for this reason but do not

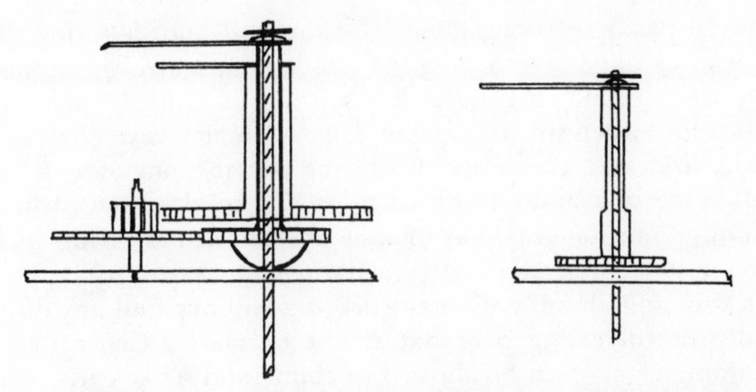

Outside plates—cannon pinion held friction-tight on centre arbor by tensioning spring from beneath or by pressed slots in pipe. Minute hand pinned on

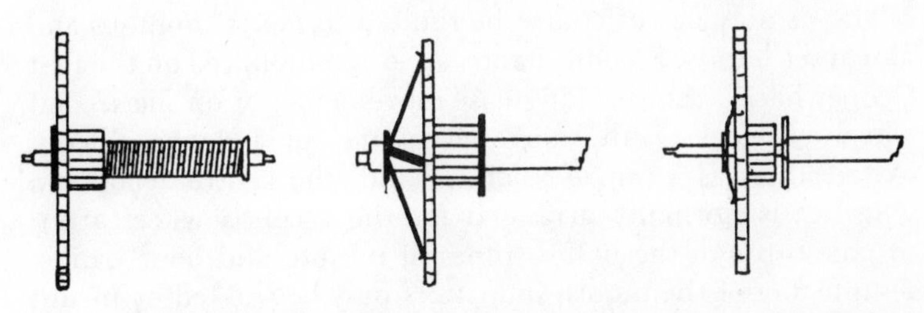

Between plates—centre wheel and pinion free on arbor but tensioned by coiled, four-arm, or flat spring. Minute hand may be pressed on

Fig. 7. Common hand setting arrangements

entirely avoid the trouble. (A similar principle applies in alarm mechanisms where the alarm set-hand should not be moved backwards.) Adjustments can sometimes be made to the hour hand independently if it is friction-tight on its pipe, but this is far from always being so, and it is as well not to adjust the position of the hour hand in this way until you are familiar with the movement concerned. The alternative is to remove the dial and to turn the large hourwheel attached to the hand until the hour is precisely shown. (Some of these points are explored more fully in Chapter 5.)

Another widespread lay belief is that a hand should not be

moved past its striking points (forward) without allowing the clock to strike fully; moving the hands should involve 'striking the clock round'. Again, it is a sound rule even if subject to variations when you know what you are doing. As we shall see, there are two quite different striking systems, one of which is self-correcting and one which must be 'struck round'; whilst chiming mechanisms may or may not be self-correcting. It is better to be safe than sorry, although once you know how striking and chiming devices work you will not find any difficulty in correcting one that is out of phase. Generally, a chiming clock must be allowed to chime and strike each hour as hands are moved, and a striking clock's hands must never be moved past twelve o'clock without letting the strike run fully. Damage or maladjustment is not always caused by infringing these guides, but there is at least a considerable risk.

Hands may not of course be limited to hours, minutes and alarm set-hands. Seconds hands are often mounted on the first (scapewheel) arbors of long-case movements, or on the second wheel of clocks with short pendulums or balance wheels. Where there is a centre seconds hand, the centre arbor and wheel it is normally arranged for the seconds wheel arbor to pass through the hollow pipes of minute and hour hands. Simple day-of-the-month indicators may be worked by means of a pin on the hour motion wheel, which will advance a 62-toothed ratchet twice in twenty-four hours, or by an intermediate motion wheel, geared to the hour wheel in a ratio of 2:1 and moving a 31-toothed ratchet once in twenty-four hours. Extensions of this indication into days of the week and months are made by means of levers pushing round star or ratchet wheels and operated by a pin at each revolution of the wheel of the preceding indication (e.g. day-of-the-week from a pin in the twenty-four hour wheel, month by a pin in the day-of-the-month wheel). For details of the many mechanisms in this area, which include calendars taking account of leap years, the reader should consult the larger manuals on horology. The operation of the various 'indicator' hands (as strike/silent, fast/slow) is fairly clear from observation, but will be mentioned also in the next two chapters, which consider in more detail the time-keeping and the striking sides of clock movements.

*Chapter Three*

# PENDULUMS, BALANCE WHEELS, ESCAPEMENTS

The heart of the mechanical clock from the point of view of time-keeping is the oscillator (pendulum or balance wheel) and the escapement which it controls. A clock is not as good as its escapement, for a high quality escapement is wasted on worn or poorly cut wheels, but certainly no clock will keep good time over a period unless the design and workmanship of escapement and oscillator are satisfactory.

*The Pendulum*

Although still found in high-grade and ornamental clocks, the pendulum has to a considerable extent been superseded as a regulator by the balance wheel and balance spring. There is some irony here in that the pendulum is younger than the balance, which, however, it went some way towards replacing in the 17th and 18th centuries. It has the merit of being robust and simple, but the disadvantage that it must hang vertically, and so is not very portable, and it occupies space, for a large and heavy pendulum is needed for good time-keeping.

Ideally, the pendulum is a weight (the bob) on a weightless line. In practice the line is a rod of straight-grained wood or of metal terminating in a screw by means of which the effective length, and therefore the timing, of the pendulum can be altered. Alternatives to the rod incorporate a series of rods of different metals, some designed to rise and others to fall, so that the pendulum is of a constant length despite expansion caused by increased temperatures, or jars or bulbs of mercury for the same purpose. Both devices may be attractive to look at, if ungainly. In small clocks they are often more ornamental than functional. Wooden and modern specialized metal pendulum rods are virtually free from the effects of temperature variation.

In the oldest clocks the top of the pendulum was screwed to an arbor in the plates and the pendulum hung from a pivot, or the arbor was sharpened to a knife-edge locating in a vee-block. Again, particularly on the Continent, the end of the rod was turned over into a hook which hung over a loop of silk thread, the latter being drawn in or let out by a screw to regulate the clock—in such cases the bob was often fixed to the rod rather than running on a screw.

For several centuries, however, the usual practice has been to suspend the pendulum from a thin suspension spring whose other end is pierced and held by a pin, or has a securing block, between the brass chops of a bracket known as the pendulum or suspension cock. The rod and spring are in some instances, especially American, made in one piece, but more often the spring is clamped into a brass endpiece and screwed to the rod (as in long-case clocks) or it has a pin inserted through it over which the top end of the rod hangs. Suspension springs are available in a variety of sizes and it is essential to replace bent or damaged springs, whose pendulums will never vibrate truly. The cock is usually secured to the movement backplate, but in Vienna Regulators and other good-quality wall clocks and long-case clocks the cock is secured to the backboard of the case independently of the movement.

Regulation by turning the screw of a pendulum rod means stopping the movement and often also entering the back of what may be a heavy clock. Therefore there have been several arrangements for regulating from the dial by means of a square or index hand, the latter having the benefit of showing how the regulation has been made from a central point. The method is generally to raise or lower a long suspension spring between close-fitting chops, thus effectively changing its length, or to raise or lower chops on a fixed suspension spring, with the same effect. Neither system is ideal, since both slightly alter the character of the suspension by varying the amount of spring (rather than of actual pendulum length) in use. The older Brocot device and the English rise-and-fall arrangement (widely adopted in economical forms elsewhere, particularly with the moving arm mounted sideways, which can distort the spring) adopt the method using fixed chops, whilst the more usual later Brocot device employs the alternative of moving chops. These French attachments are not easy to illustrate or explain, but they are readily recognized and understood in practice (Fig. 8).

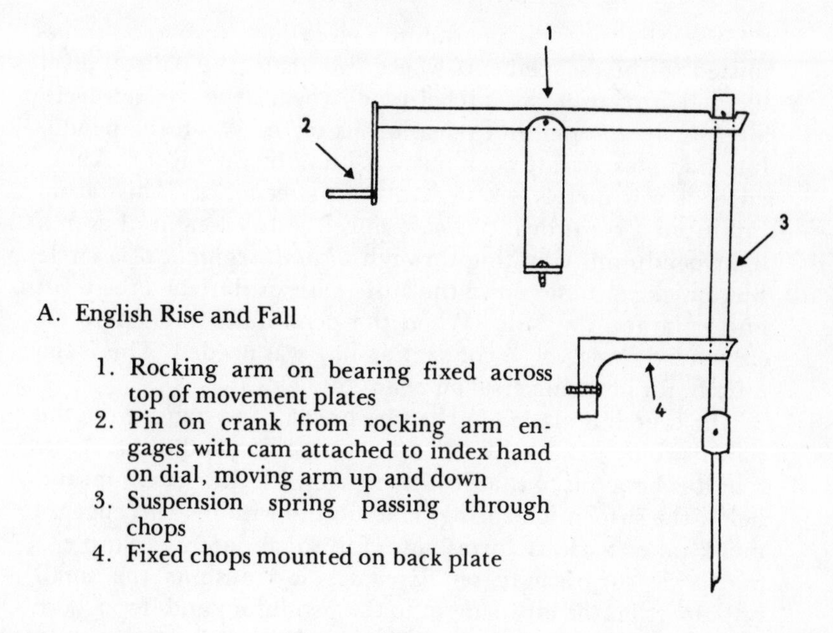


A. English Rise and Fall

1. Rocking arm on bearing fixed across top of movement plates
2. Pin on crank from rocking arm engages with cam attached to index hand on dial, moving arm up and down
3. Suspension spring passing through chops
4. Fixed chops mounted on back plate

B. Brocot

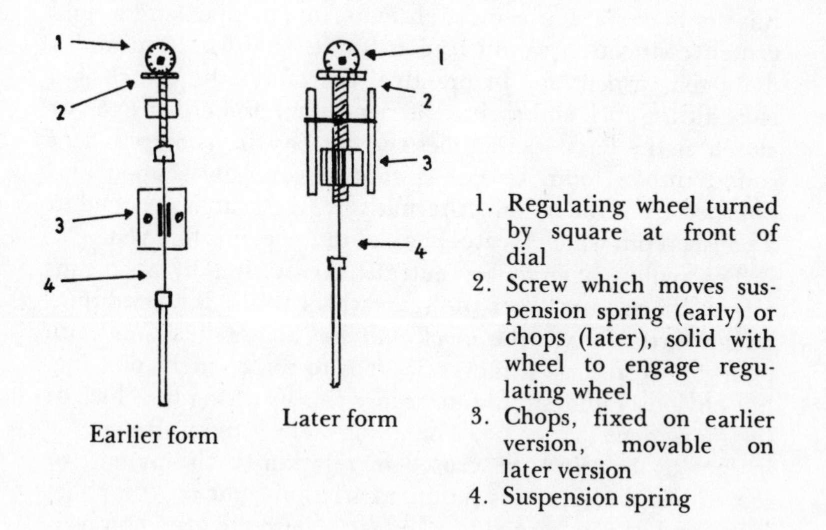


1. Regulating wheel turned by square at front of dial
2. Screw which moves suspension spring (early) or chops (later), solid with wheel to engage regulating wheel
3. Chops, fixed on earlier version, movable on later version
4. Suspension spring

Fig. 8. Regulating pendulum suspensions

As we shall see, the pendulum's vibrations have to be transmitted to the escapement, where two arms (the pallets) pivot to and fro over a ratchet-type wheel (the scapewheel), alternately holding and releasing its teeth. When the pendulum had no separate cock but was hung by the pivot or knife-edge, it was merely screwed to the pallet arbor. This set-up produced a great deal of wear, since the full weight of even a light pendulum vibrating through a small segment of a circle very quickly ground down the knife-edge or flattened the pivot and enlarged the hole. When the pendulum was separately suspended, however, a connecting link was needed. This is the crutch, for long universal on pendulum clocks.

The crutch is made of light wire or strip and hangs from the pallet arbor much as did the older type of pendulum. At its foot it is bent out so that it will engage with the pendulum rod below the suspension spring, and vibrate with it. This engagement takes various forms and for so apparently simple a matter is surprisingly sensitive, for it transmits the small impulse from the escapement to the pendulum and if too loose will waste power or, if too tight, will bind. Whilst the clock must be set horizontally (or adjusted for being out of the level) it is not material if it leans slightly to front or rear, and a rigid crutch connection would lead to unnecessary friction and a damaged suspension. In practice, therefore, the crutch is a close-fitting fork embracing the pendulum and either open or closed at the back. Sometimes the crutch wire is merely bent round into a loop, or else a fork is carefully shaped and polished from the metal. Alternatively, the crutch may end in a polished pin which locates in a slot in the pendulum rod.

The pallets have to be central (usually, but by no means always, horizontal) over the scapewheel whilst the pendulum hangs vertically, or the clock will be 'out of beat' and run poorly if at all (*see* Chapter 8). It is to some extent possible, but obviously undesirable, to secure this by tilting the clock or the movement in its case. The adjustment is most often made in fact by bending the crutch in relation to the pallets, or sometimes the crutch is mounted friction-tight on the pallet arbor and can be turned to adjust it. There are also, however, adjustment facilities built into some crutches, especially those of Vienna Regulators (weight-driven wall clocks with glazed fronts and brass-covered weights) and some good modern mantel clocks. In the main there are two forms of device; either the crutch pin is located in a screw which may be turned to move the pin towards one side or other of the centre-line of the crutch, or the crutch is in two parts, connected friction-tight, one of which can be set more or less out of the straight in relation to the other.

*The Balance Wheel*

The balance wheel has likewise to be connected through pallets to the gear train, and the all but universal modern method is by means of a pivoted lever, at the other end of which are the pallets. As the pallets move from side to side, they intersect the circumference of a circle on which the scape-wheel teeth are placed. The connection of lever to balance varies considerably but is generally by means of a pin moving in a forked notch at the lever end. The pin (the impulse pin) may be mounted in a steel disc (the roller) driven on to the balance arbor (the balance staff), this disc having a notch immediately in front of the pin. A point or pin at the tip of the lever passes into this notch so that the wheel can move the lever only when roller and lever are correctly aligned. The impulse pin then engages with the lever fork so that the lever is moved and also so that impulse is passed by the lever to the balance. Alternatively, there is a double roller, in which two steel discs are linked, one holding the impulse pin and the other having the passing notch. In the cheaper pin-pallet escapement there is often no roller as such, the passing bay being formed by a flat in the balance staff and the impulse pin being a steel pin driven into a balance wheel crossing in front of the bay. A lever with an inner and an outer fork is then used, the horns being so designed as to prevent the lever from passing except when the passing bay is aligned with it. (Some lever and roller arrangements are illustrated in Fig. 9.) There are of course other balance wheel escapements and in one, common in 19th century clocks, the cylinder escapement employs no crutch or lever, the balance staff engaging directly with the scapewheel. More will be said of this later.

Modern escapements are 'detached'—for most of the vibration of the balance it revolves without being in contact with

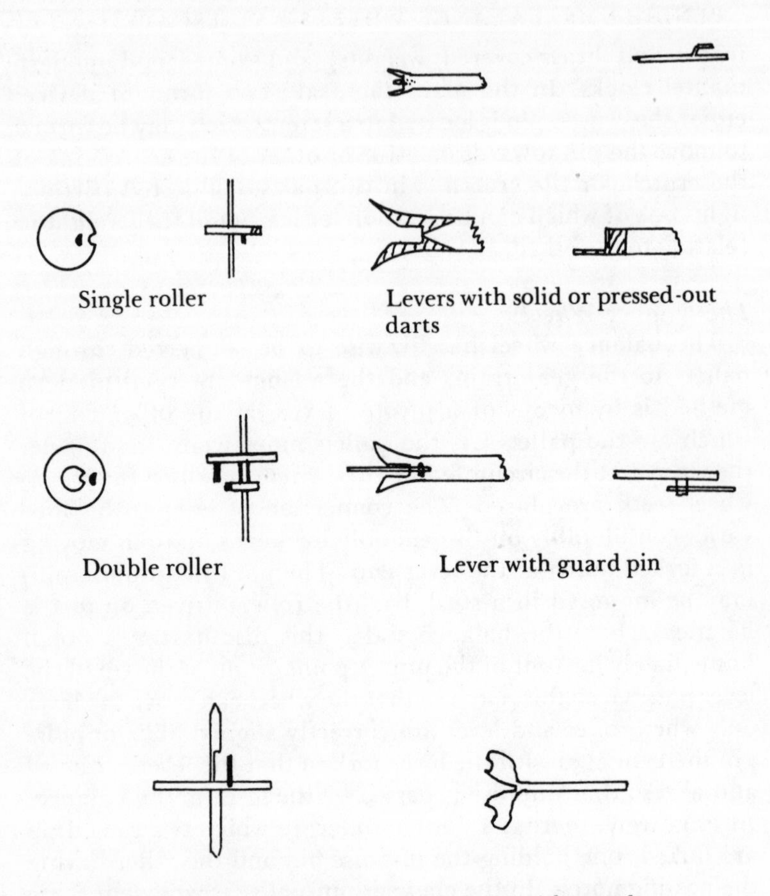


Fig. 9. Balance roller and lever arrangements

the train, and as a result it has to be returned to the central position by a balance spring, for it cannot be returned by the action of the pallets alone as could old frictional escapements where the pallets were solid with the staff and constantly in contact with scapewheel teeth. The balance spring is pinned at its inner end into a brass washer (collet) which is driven on to the balance staff in such a way that the wheel returns to a central position after swinging. The other end is usually pinned into a small brass block mounted on the top bearing for the balance (the balance cock). The timing is finely regulated by an index terminating in two close pins, or a pin and a block, on each side of the balance spring, and these are moved so as to lengthen or shorten the effective length of the balance spring. If regulation is needed beyond the scope of the index, more or less balance spring must be pinned to the balance cock, and the spring will have to be turned on the staff so that the central alignment, which will be impaired by this adjustment, is restored. The balance is thus 'set in beat' again, with the impulse pin in the lever fork and the pallets central to the scapewheel (*see* Chapter 8). Good quality balance wheels may have heavy screws in their rims; turning these out or in varies the distribution of mass so that the wheel can be truly balanced, and the screws are also used for regulation. It is not, however, normally necessary to adjust these on clock balances and they are best left alone.

*Pendulum Escapements*

The basic pendulum escapements of modern times are the anchor or recoil and the dead-beat, of each of which there are several forms. The outlines are illustrated in Fig. 10. In both escapements, pallets are pivoted over a scapewheel, and on the pallet arbor is fitted the crutch, so that the pallets rock sideways, alternately locking and releasing the teeth of the wheel. The critical point in their operation is the shape of the wheel teeth and pallet nibs, and various combinations are found according to date, whether the clock has a long or short pendulum (the latter with a larger arc of vibration), and for economic reasons, which dictate that pallets of cheaper modern clocks are made of bent steel strip with the edges sharpened to an angle, rather than of a solid steel block. Similar factors also govern how many teeth are embraced by the pallets.

The essential difference between the two forms of escapement is that in the recoil escapement the continued swing of the pendulum after a wheel tooth is locked forces that tooth to move backwards whereas, as the name implies, in the deadbeat escapement the scapewheel does not move once it has been locked, and in practice the pendulum is maintained at a

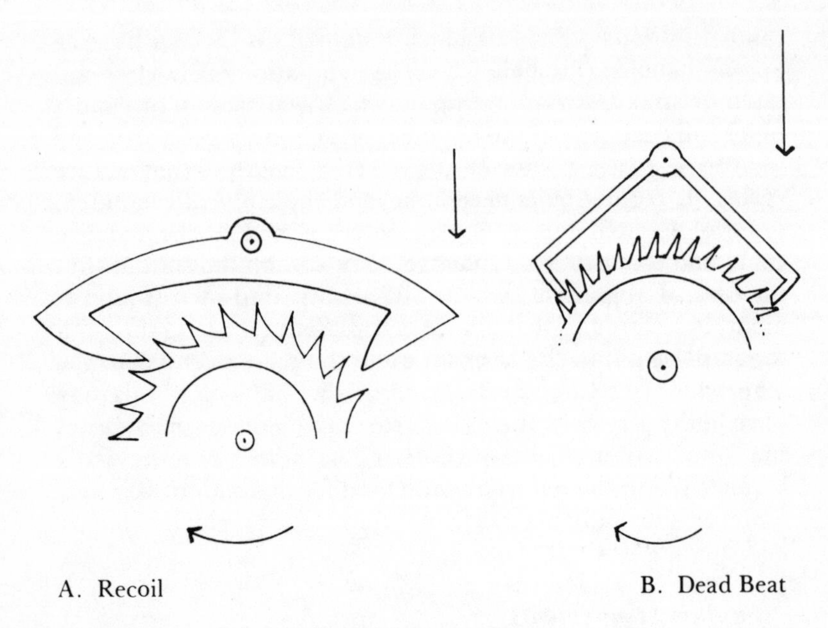


Fig. 10. Escapements

smaller arc. This latter is the finer, more accurate escapement, but it is less robust and it will run less well in adverse conditions of dirt, wear, and inconstant power supply. There are also compromise forms in which there is slight recoil, or recoil on only one of the two pallets. In the true recoil escapement, the released scapewheel tooth falls on to the curved face of the pallet, which both holds it whilst the pendulum swings on, and then the tooth imparts impulse to the pendulum on its return swing. The true dead-beat pallets, by contrast, have two distinct surfaces, the first (at the 'side') serving to lock the wheel teeth, and the second (at the 'foot') being the impulse face along which the escaping tooth slides, imparting impulse as it goes. It will be noticed that the recoil pallets are planted close to the scapewheel and that the teeth engage with them fairly deeply, whilst the dead-beat pallet arbor is set further above the wheel and locking on the corner of the pallet is as shallow as it can be with safety. These distances are critical in escapement design.

There has to be an element of freedom in these as in other escapements, or both pallets would be in contact with wheel teeth at the same time and the escapement would jam. This freedom is known as 'drop', and it is kept as small as possible. It can be observed as the gap between one tooth and pallet as the opposite tooth is on the point of being released by its pallet, and it should be tested and found to be the same for every tooth of the scapewheel. Adjusting the drops, so that maximum impulse is given but no jamming occurs, is one of the main jobs in servicing an escapement. Another matter requiring attention is the depth of the escapement – how close the pallets are to the wheel. This is often adjustable in modern clocks by means of slotted screwholes in the rear pendulum and pallet cock, and sometimes also with an eccentric screw pivot hole in the front plate. In older clocks adjustment here may be more difficult. These points are discussed in more detail in Chapter 7.

Fig. 11. Brocot escapement

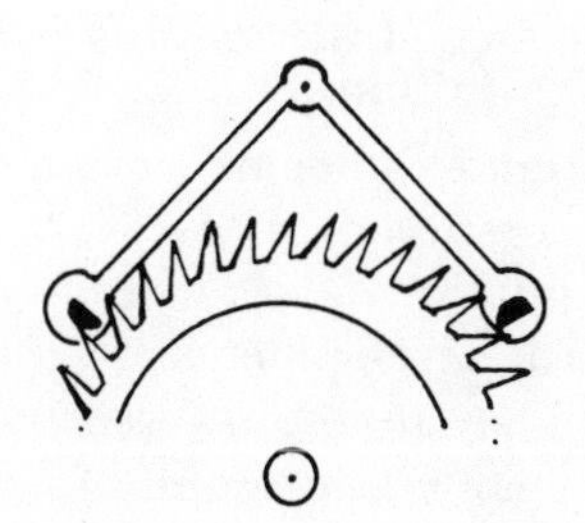

Two developments of these escapements are found in clocks of a certain age and also on those more modern, usually in dead-beat form. These are Brocot's pin-pallet escapement and the Vienna regulator and Four hundred day (Anniversary) clock escapement. The former (Fig. 11) is particularly common on French clocks, or clocks of French pattern, of the latter part of the 19th century, when visible escapements were popular. Here the train is arranged so that the scapewheel is mounted on a cock in front of the recessed centre of the dial, and the pallets hang down from the rim above, being pivoted at the front in the rear of the dial plate. The pallets are short semi-circular rods, and may be of steel or of attractive red stone or semi-precious material. The escapement is very reliable, but the teeth are fragile and often have to be straightened, and the pallets themselves are not always correctly placed (with the flat faces radial to the scapewheel arbor). Locking takes place at the apex of the pallet curve; if the teeth touch the flat sides of the pallets or if they lock imperfectly on the lower curved impulse surface the clock will certainly stop. The wheel teeth vary considerably in shape. The best shape seems to be with radial faces, but sometimes the face is undercut to ensure that only the point engages, and then the teeth become very weak. A curved back has only the advantage of strength and plays no part in the working of the escapement. If the face of the tooth is not radial, but sloped slightly forwards, recoil is introduced, although this is of little benefit in the fine and rather delicate movements for which this escapement was generally used.

The Vienna regulator and Anniversary clock escapements are straightforward developments of the classic dead-beat pattern (Fig. 12). They almost always employ steel strip pallet pieces, these being slotted into solid brass arms and clamped there by screws so that they can be adjusted in and out. As a result, the adjustment of depth and drop is simplified, especially as an eccentric screw pivot hole is normally used. Many of these pallets are shaped at both ends, so that to rectify wear it may only be necessary to reverse the pallet in its slot and so offer up a brand new surface—few are yet old enough for the second surface to be much worn. This is fortunate, for the cutting of these pallets from hard steel of the precise curve

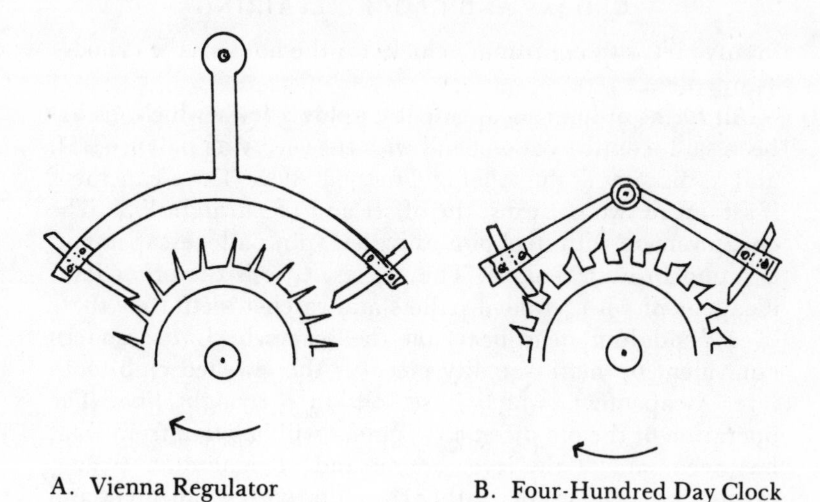

A. Vienna Regulator B. Four-Hundred Day Clock

Fig. 12. Escapements

(the circumference of a circle centred on the pallet arbor) is no easy matter. The wheel teeth vary much as those of the Brocot escapement, except that recoil is not often introduced. It should be noted that the Vienna escapement locks shallow, like the traditional dead-beat, but the Four hundred day escapement locks deep, the rotating torsion pendulum having a substantial overswing after locking has taken place. These latter clocks can try the patience of the most experienced adjuster and shallow locking should be one source of trouble which is eliminated at the outset.

These are the principal types of pendulum escapement found in the commoner domestic clocks of today and yester-year. There are many variations but, if the principles are grasped and some experience is gained, they are not likely to cause insuperable problems.

*Balance Wheel Escapements*

The balance wheel escapement now generally used is the lever escapement, in several forms. There are other older escapements, some hardly more than experimental although produced in quite large numbers, but the only other one commonly met with in ordinary clocks for the house is the cylinder escapement.

All forms of lever escapement employ a lever which, as has been said, connects at one end with the revolving balance staff and terminates at the other end in the pallets. The escapement is set out in two patterns, the offset and the straight-line. The cheap version with steel pins as pallets (pin-pallet escapement) is found in both designs. The earliest type is the offset lever escapement with jewelled pallets and ratchet teeth (like those of a pendulum dead-beat) on the scapewheel. Its modern equivalent on better-quality pieces is the jewelled club-tooth lever escapement, which is set out in a straight line. The operation of the old offset escapement will be plain from what has been said of the lever action and of pendulum escapements. We will consider rather the club-tooth escapement and the pin-pallet escapement (Fig. 13).

The action of the club-tooth escapement is comparable with that of the dead-beat pendulum escapement in that the locking occurs on the side corner of the pallet and the impulse on the angled foot, but there is also slight recoil. The locking is as shallow as can be secure but, after it, the side of the pallet is so angled that the wheel tooth continues to pull the pallet downwards. This action is called 'draw' and it ensures that the balance can swing detached from the escapement, with the lever held clear of the balance staff, until the return vibration releases the tooth. The 'club' profile of the tooth is such that its mating with the foot of the pallet conveys the desired impulse. There are usually fifteen of these teeth and the pallets embrace three teeth with, in addition, a tiny clearance for drop. The pallet stones are mounted in slots with shellac and can be moved in and out after gentle warming, though this is a very delicate operation. There is also sufficient room in the slots to permit lateral moving for the adjustment of draw.

Except in layout, the pin-pallet escapement, which is normal in cheap alarm movements and spring-driven movement of all sorts (including those wound by electricity), does not differ in its offset and straight-line forms. The pallets are steel pins, as thin as is compatible with strength, driven into a lever, which is usually brass. As has already been mentioned, the other end of the lever is often a double fork operating on a

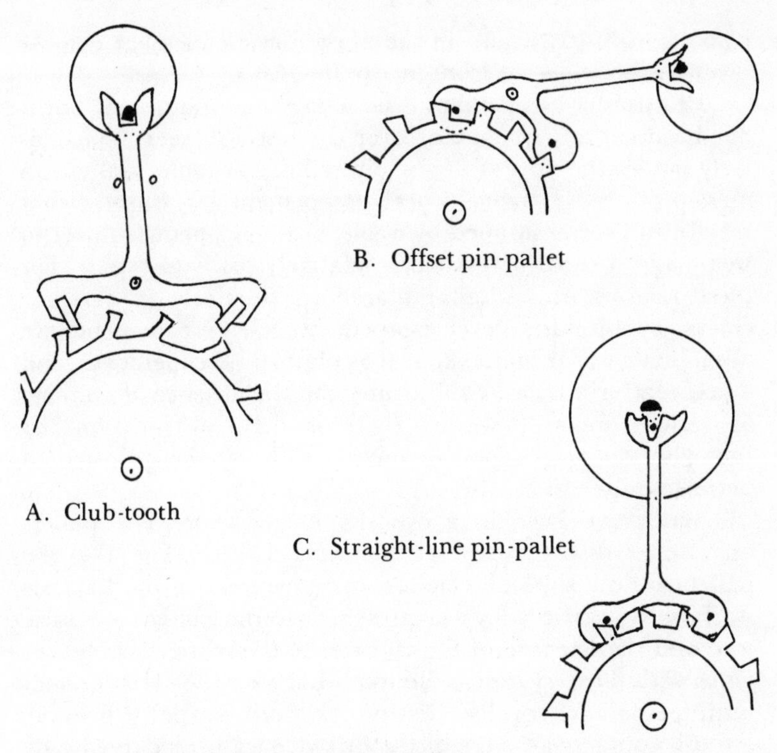


Fig. 13. Lever escapements. For details of levers and rollers, *see* Fig. 9

balance staff with no roller, but the more traditional roller systems are also used although sometimes on modern movements these are made of synthetic material. In the double fork there is no shaped pin (guard pin) or pointed tip on the lever to enter the passing bay, for the balance staff itself lodges in one side of the outer fork and so prevents the lever from crossing unless aligned. The wheel teeth are of a distinctive stub-shape, and it must be arranged that pallets lock on the front face of these teeth, not on the angled impulse surface. Sometimes there are pins or raised sections to prevent the lever from moving too far (as in most lever escapements), and sometimes there are no such banking pins, matters being adjusted so that the pallets bank at the feet of the scapewheel teeth. The draw compels the lever to remain banked whilst the balance completes its vibration, and in the pin-pallet escapement draw is given by the undercut front face of the teeth.

The cylinder escapement is no longer manufactured, but it has been extremely popular over the last 150 years, particularly before the advent of the pin-pallet as the normal cheap mass-produced balance wheel escapement. It is somewhat fragile and easily stopped by a jolt, as well as being subject to wear, and it is also, nowadays, relatively costly to repair. For these reasons it is often replaced, particularly in carriage clocks, by a modern lever escapement. Such escapements are often made as sub-units, known as platform escapements, and a new platform is easily substituted. In the absence of damage or serious wear, however, there is little justification for removing an original escapement which can be adjusted to perform quite efficiently.

There is no lever in a cylinder escapement. The pallets operate on distinctive raised scapewheel teeth (Fig. 14); the pallets are the sides of a hollow open cylinder in the balance staff which, as it vibrates, presents alternately the cylinder wall and the opening of the cylinder to the scapewheel tooth. Once within the cylinder, the tooth butts on the cylinder wall until the balance revolves, letting the tooth escape. Whenever a tooth contacts an edge of the cylinder wall, its curved face presses against the wall and imparts impulse to the balance. Thus the escapement is highly frictional, the cylinder and balance being in virtually continuous contact with the scapewheel. It is clear that if the cylinder revolves too far it will catch the underside of a tooth and cause damage. Therefore the balance is only allowed to make half a revolution in each direction, its vibration being limited by a banking pin fixed to its rim; this pin contacts another fixed in the back of the balance cock should the balance attempt to over-swing. The drops are of course between the tip of the tooth and the outside wall of the cylinder, and again between the tip and the inside wall within the cylinder; the tooth must have slight freedom within the cylinder, and equally the cylinder must be slightly smaller than the space between two teeth or it will jam.

This completes our survey of the time-keeping elements of a non-electric clock, though more will be said of their repair and adjustment in Chapter 8. Many versions of these basic escape-

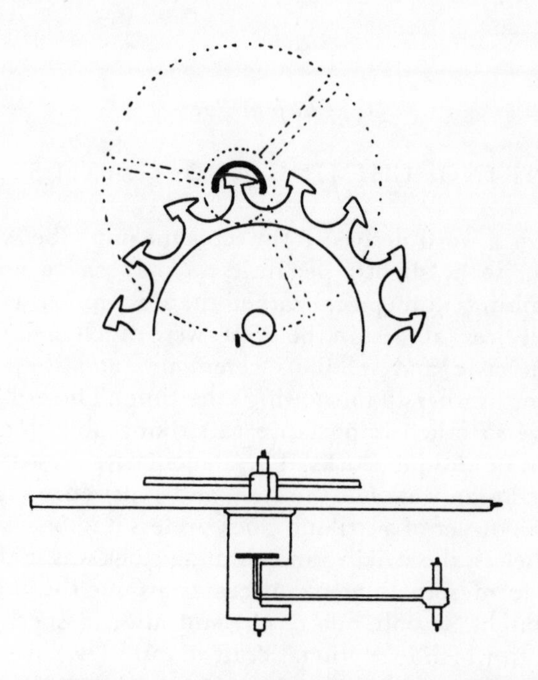

Fig. 14. Cylinder escapement (showing tooth in cylinder)

ments, and also of others, will be found in domestic clocks, particularly in older ones, but their operation and adjustment can usually be deduced without difficulty from familiarity with the commonest types.

*Chapter Four*

# STRIKING, CHIMING AND ALARM SYSTEMS

'Clock' is a word derived from roots meaning 'bell', so that if we want to be slightly pedantic we shall call a non-striking mechanism a 'timepiece' rather than a 'clock'. The earliest clock-devices, at least in the West, were much used to indicate time in civic and religious ceremony, and they relied on sounding, rather than showing, the time. These days, we do not give so much importance to striking and chiming, save perhaps in antique clocks. These ancillaries are less generally provided than was once the case, and experience suggests that many an owner of a striking clock prefers it to be set to 'silent'. Nonetheless, the striking and chiming clock was in the past, in simple form, something of a necessity around the house (where there might be only one clock) and also, in special cases, a luxury to which exceptional craftsmanship was devoted in its making and decoration. Moreover, many clocks stop completely when the sounding mechanism is broken or improperly silenced, and if you have come by a striking or chiming clock you will want it in good order throughout. Thus although, as was said in Chapter 1, striking and chiming are in the nature of attachments let off by the principal going train, there is a strong case for giving these no less attention than is given to, for example, the escapement. The same applies to alarm work although, if it is not specially required, it is unlikely to stop the clock through not working correctly.

'Striking' is normally used of the number of strokes sounded at the hour, and 'chiming' of a proportional tune sounded at the quarters; the functions are generally performed by separate trains, but there is also 'quarter-striking' in which two or three notes are struck several times according to the quarter of the hour, and this is run from the hour striking train.

There are two quite different systems of striking, which are applied also, often in combination, to chiming. Broadly, the older, 'countwheel' system has the merits of strength and cheapness, whereas the later 'rack' system (invented in the 17th century) has the advantage that, if correctly set up, the striking will always correspond to what is shown on the dial. The important aspects of striking and chiming mechanisms may be considered as regulation (striking the correct number of blows), letting off (starting the train), and locking (stopping the train). These factors have to be considered for each system. There are also features common in both systems, being 'warning' (not always present) and the operation of the bell or gong hammer. So far as the latter is concerned, the hammer is usually a pivoted lever, the longer section terminating in a metal or leather-tipped head, and the short 'tail' (at an angle to the hammer) coming into the path of pins placed round a 'pinwheel' (or 'hammer wheel') between the plates. The pins raise the hammer tail until it drops off and the head strikes the gong or bell. The hammer is usually, but not always, sprung to assist the blow, and the head is placed clear of the bell so that it does not jangle but strikes the bell sharply and moves away again. The arrangements of springing and hammer-stop vary, but they will be clear from observation. In modern clocks a starwheel is used instead of a pinwheel, for it is both stronger and cheaper to produce. Chiming hammers operate similarly, but the pinwheel is usually replaced by a series of pinwheels or a pin-barrel, so that there is a circle of spaced pins for each hammer. In some arrangements, including the common cuckoo clock, two or three hammers may work off one pinwheel and be used for striking quarters.

*Countwheel Striking*

In the countwheel system, regulation is by the countwheel or 'locking plate'—the latter something of a misnomer since the wheel does not actually lock the train but merely indicates where it may be locked. The countwheel is a large wheel with raised sections around the edge corresponding to the time to be taken up at the striking of each hour. As 78 blows in twelve hours have to be struck in an hour-striking clock, the appropriate countwheel is based on a circle divided into 78 segments. The raised section corresponding to two o'clock is one division, allowing two strokes, there is then a slot of one division where the train is locked, and then three o'clock is represented by a raised section of two divisions, and so on. The

French countwheel, and many others, are based on 90 divisions, allowing for a single blow to be struck at each half-hour as well as for hour striking; the half hour is given by having each slot two divisions wide (Fig. 15). In the older clocks, and in French clocks throughout the 19th century, the countwheel was usually mounted outside, on the back-plate of the movement (*see* Plate 2), either (in an eight day clock) on the extended arbor of the wheel next to the mainspring barrel, or (in thirty-hour clocks) on a stud and geared into the mainspring or weight barrel. In some American clocks the internal countwheel has gear teeth cut in its sections and is advanced by a pin or extended pinion leaf on the locking or hoop wheel (mentioned below).

In the following discussion it may be as well to refer back to Fig. 2 (p. 20), which illustrates the arrangement of wheels in a striking train. Old English and some later clocks lock the train by means of a hoop wheel, a disc with a notch in it into which drops a detent (the locking piece), which has a counterpart (the countwheel detent) able to drop into the slots of the countwheel. Thus the countwheel and hoop wheel revolve at different speeds, but they are so geared that every so often the notch of the hoopwheel and a slot in the countwheel will coincide, whereupon the detents fall and the train is locked by the locking piece in the hoop wheel notch, Fig. 16. French clocks and many modern clocks, including chiming clocks, employ a locking wheel instead; this has a protruding pin which engages similarly with the locking piece when a slot in the countwheel permits it to fall (*see* Plate 2). Geared between these two wheels in the train is the pinwheel (which in the old thirty-hour long-case movement is the greatwheel carrying the chain pulley). Reflection will show that the gearing between these three wheels has to be precise so that at appropriate intervals the slots of the countwheel and the notch of the hoop wheel (or pin of the locking wheel) do in fact coincide, and so that, for each revolution of the locking or hoop wheel one blow is struck by the hammer operated by the pinwheel.

The striking train is let off by a lifting piece which is not on the same arbor as the countwheel detent and locking piece, but which raises them by means of a pin or attached lever and so releases the train (Fig. 17). The lifting piece takes various

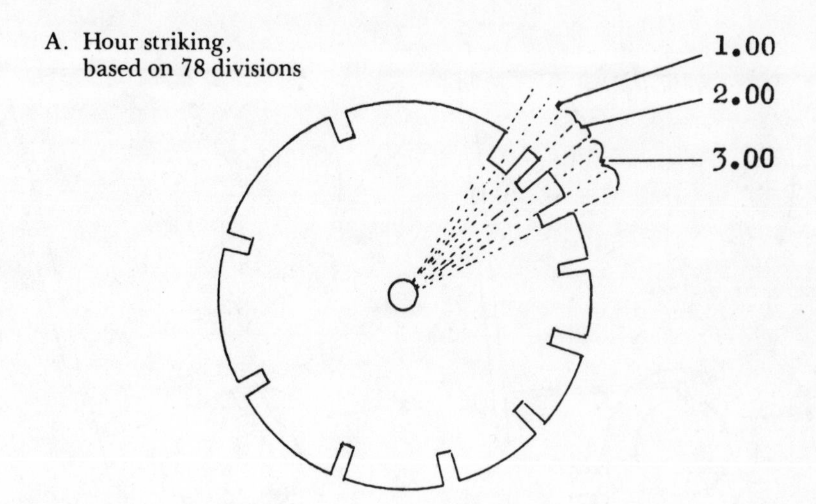


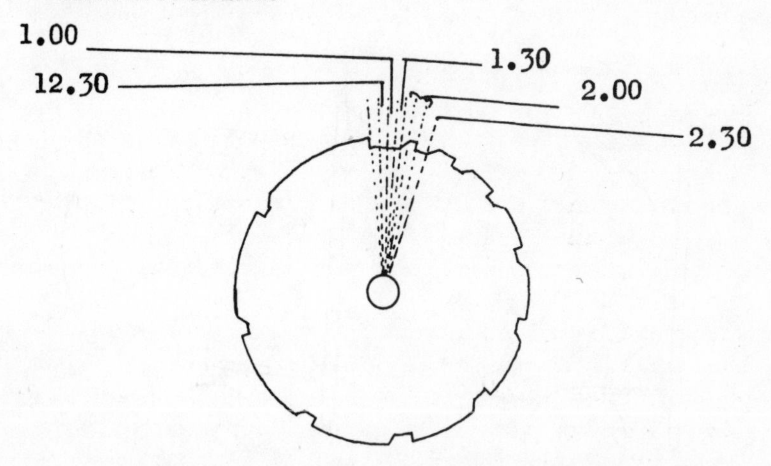


Fig. 15. Countwheels for hour and hour-plus-half-hour striking (*See* Plate 2)

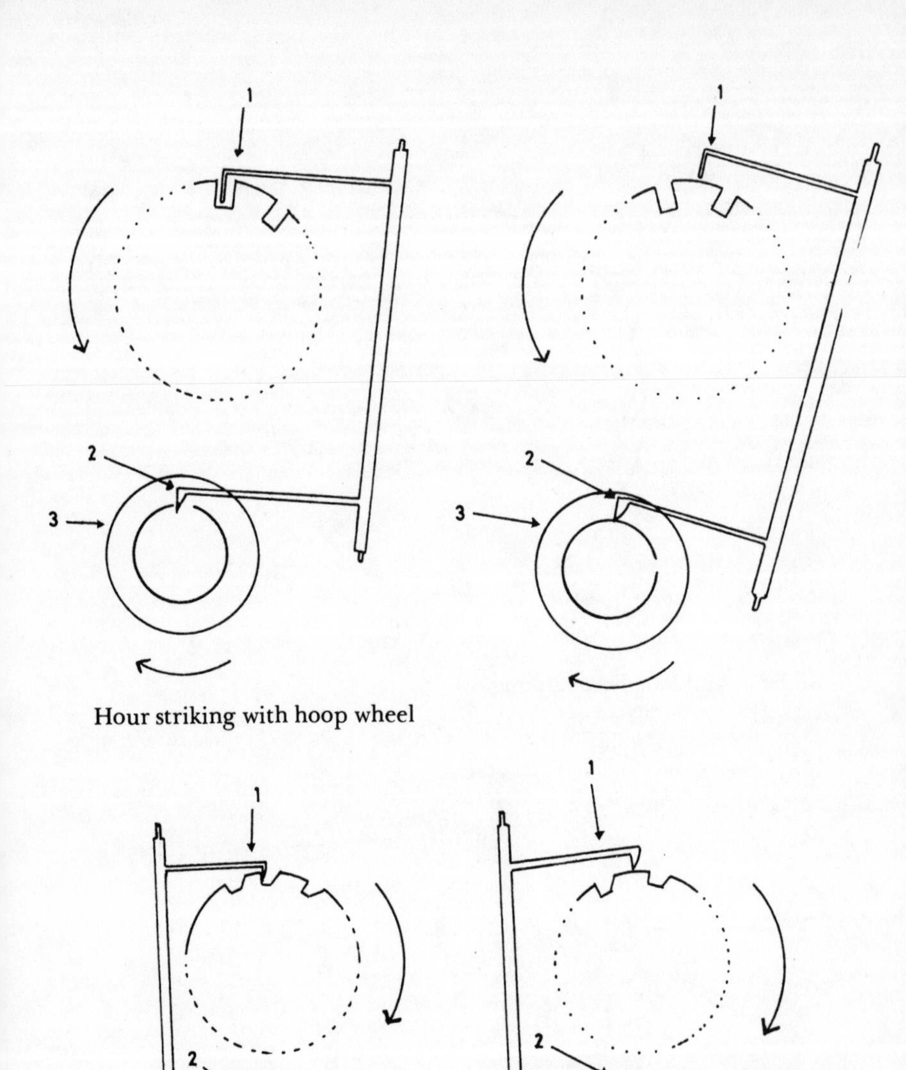


Hour striking with hoop wheel



Hour and half-hour striking with locking wheel

1. Countwheel detent
2. Locking piece
3. Hoop wheel with notch
4. Locking wheel with locking pin

Fig. 16. Locking of countwheel striking (*see* Plate 2)

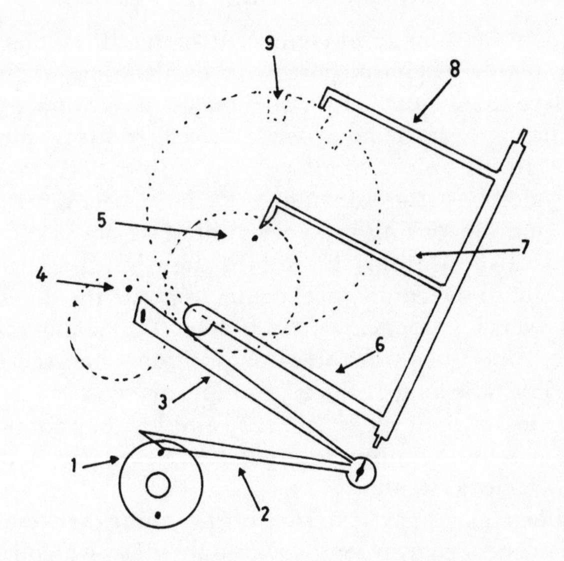


1. Cannon pinion or minute wheel with lifting pin
2. Lifting piece
3. Warning piece (may be internal, on same arbor as lifting piece
4. Warning wheel and warning pin
5. Locking wheel and locking pin
6. Pin or lever connecting warning piece and locking piece arbor. (An internal warning piece may merely have a projection which presses up the locking piece from below)
7. Locking piece
8. Countwheel detent
9. Countwheel

(Nos. 1–3, and sometimes 6, are usually on the front plate. 8 and 9 may be outside the back plate. Remaining parts are between plates)

Fig. 17. General system for release of countwheel striking (*see* Plate 2)

forms but is always basically a lever across the front plate, and its end is raised by a pin in the cannon pinion (centre arbor wheel) or the minute wheel. However, though released, the train only runs for a second, since it is immediately obstructed by the coincidence of a lever attached to the lifting piece (or on its arbor) and a pin on the next wheel in the train (the wheel preceding the fly). This is the warning wheel with its warning pin, and the lever connected to the lifting piece is the warning piece. Their purpose is to ensure that, though the train is free to run just before the hour, it does not actually do so until precisely at the hour, when the raised lifting piece falls from the lifting pin, bringing down with it the obstructing warning piece. At the warning, the countwheel detent leaves the slot, and the locking piece rises out of its slot or off its pin, so that once the warning piece falls the train must run until the next slot of the countwheel comes opposite the detent. The warning wheel is so geared that, like the locking or hoop wheel, it always stops in the same position when the train is locked. The 'warning', which is found on most striking and chiming clocks whatever system they employ, but not on all, is some or all of the preparatory noise heard three minutes before such clocks sound.

It will be clear that the nature of the countwheel train is to run in a sequence governed by the countwheel which is a part of that train and that therefore there can be no question of automatically correcting striking or of moving the hands back to obtain a correction. If the clock strikes three when it should strike four, the train will have to be run and the clock struck an extra hour for the striking to catch up with the hands—a wire or string is often provided so that this correction can be made from the front or side. Although safety devices are sometimes incorporated, it is not advisable to push the hands back to three in the search for correction. If, on the other hand, the striking is ahead of the going (for example striking five at four o'clock), then either the striking must be taken right round, or the hands must be moved in such a way that the clock does not strike as they pass five o'clock. Thus it is possible to move the hands quickly to six without the clock striking five (though it is not very advisable, and will not work if more than one hour's advance is involved), or the striking train will have to be manually arrested while the hands are advanced past five, to six o'clock. All this is less involved than it sounds and indeed striking and going trains should not get out of sequence if the clock is sensibly stopped before it runs down (during holidays, for instance) and the hands are not advanced without permitting striking to take place at each hour (and half hour). It does happen, however, that movements become worn in their locking and will strike at the slightest pretext, particularly as they are being wound. Then they may need to be struck round with the lever or cord, or by raising the detent on an external countwheel on the back plate. Subject to these occasional irritations, the countwheel system is very sturdy and reliable.

*Rack Striking*

The alternative, the rack system, has the convenience that it cannot go out of phase (provided it is in good order), because the regulation is controlled not by part of the striking train itself but by part of the going train which is directly connected to the position of the hands. This is the 'snail', a stepped wheel whose highest raised section produces one stroke on the gong and whose lowest portion produces twelve strokes. Sometimes the steps of the snail are not cut, and it appears as a continuous cam, but there is no difference in running, provided that proper care has been taken in positioning the snail in the first place. The snail turns in twelve hours and in cheaper clocks is fitted to the hour wheel (*see* Plates 3–6) so that, unless the hour hand is loose, its position must correspond to that shown by the hour hand. With superior quality clocks, particularly old English bracket clocks and carriage clocks, however, the snail is mounted on a starwheel revolving on a stud (*see* Plate 7), and is advanced hourly by a pin on top of the cannon pinion (Fig. 18). This means that the snail position can be adjusted to match the time of day when the clock is set up, and it also means that the snail flicks into position rather than changing gradually during the hour; as a result, it is suited to repeaters, where the previous hour should not be struck until the last two minutes before the new one. (That is, a repeater should strike eleven when its button is pressed even though it is almost twelve o'clock—such is the custom, and it is essential to quarter repeaters, but the adjuster may prefer to have hourly repeaters striking to the nearest hour and set the snail to move at half past the hour.) The disadvantage of the arrangement is of course that the snail may be moved independently—in particular, the hour wheel can scrape against it and produce an elusive inconsistency if the snail stud is crooked or bent.

The snail is connected to the striking train by means of a

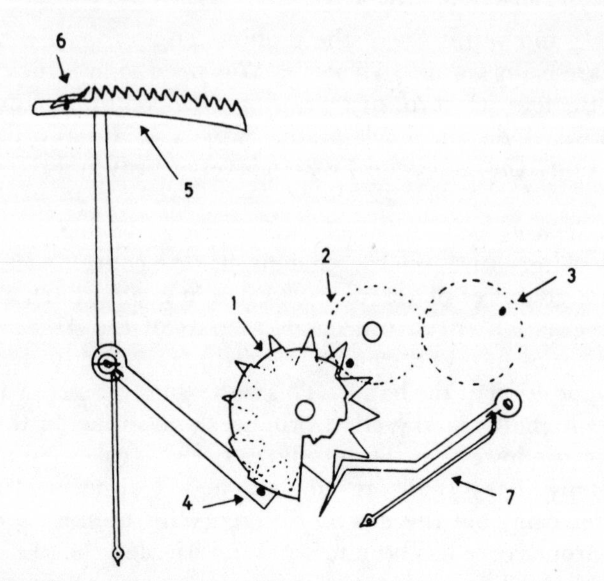


A. Old English type

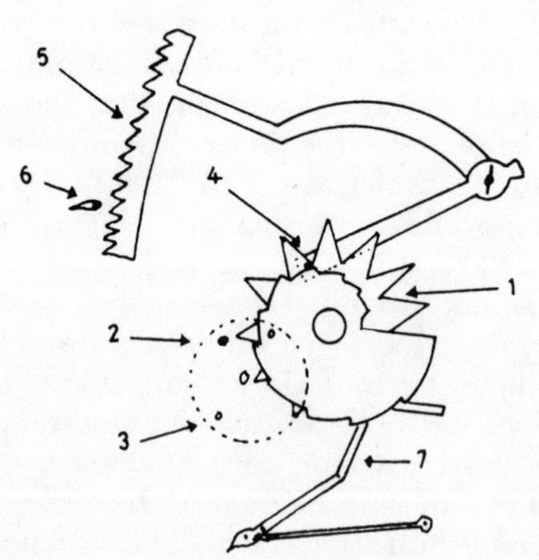


B. French type

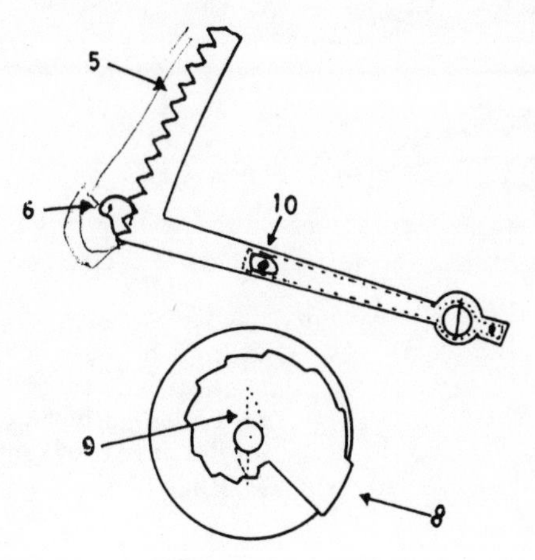


C. Modern type

1. Snail fixed to starwheel
2. Pin on cannon pinion advances starwheel hourly
3. Lifting pin on cannon pinion or minute wheel
4. Rack tail with sprung pin to engage snail
5. Rack
6. Gathering pallet
7. Jumper and spring to restrain starwheel
8. Snail fixed to hour wheel
9. Lifting cams on centre arbor
10. Integral rack tail, sprung pin in middle of rack

Fig. 18. Arrangement of rack and snail (*see* Plates 3 and 4)

rack, a pivoted arm whose lower end, the rack tail, falls by gravity or a spring (or both) on to one of the segments of the snail. The upper end of the rack is of course toothed and into its teeth meshes a pin or tooth on the extended arbor of the locking wheel in the striking train. This tooth is known as the gathering pallet since, as it revolves against the rack, it gathers it up, tooth by tooth, until the foot of the rack rests on a hook, the rack hook, which falls into place below it. This rack hook, in the French and in many modern systems, is either on the same arbor as the locking piece which locks the train (as in the countwheel system), or carries an extension back into the train

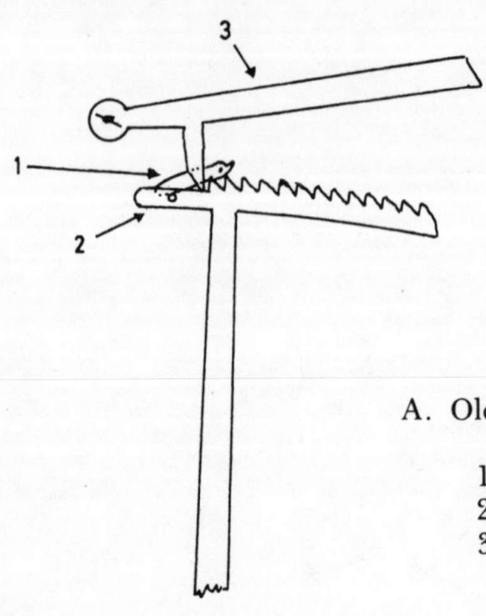


A. Old English (pallet locking)

1. Tailed gathering pallet
2. Pin at rear of rack
3. Rack hook

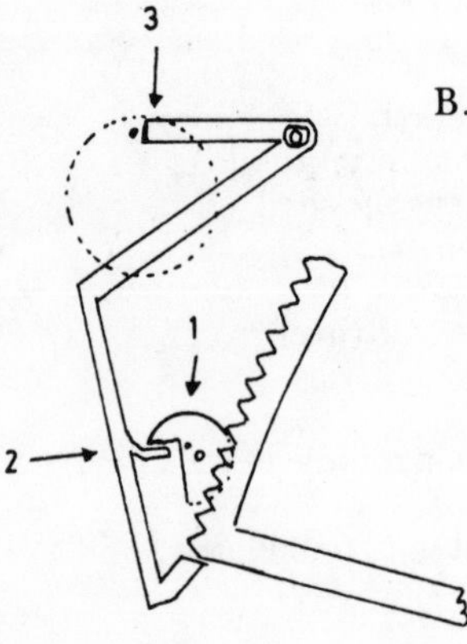


B. Modern types

1. Gathering pallet (pin). Face of cam may lock train
2. Rack hook
3. Locking piece (attached to rack hook) and locking wheel between plates. Locking wheel is second wheel doubling as warning wheel

which has the same function. In the old English arrangement (*see* Plate 3), the train was locked not by a separate locking piece but by a long tail on the gathering pallet contacting a pin at the top end of the rack, thus preventing further rotation—the same system is employed in some French movements. In many modern devices, the gathering pallet is in the form of a pin on a cam, and this cam has a straight edge which locks on a projection of the rack hook (Fig. 19). The purpose of the cam is to move the rack hook in and out of the rack teeth as they are gathered.

Whatever the locking arrangement, when the train is let off,

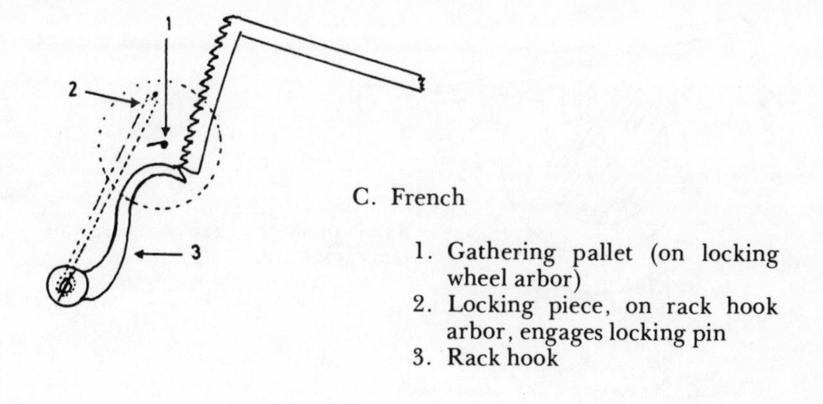


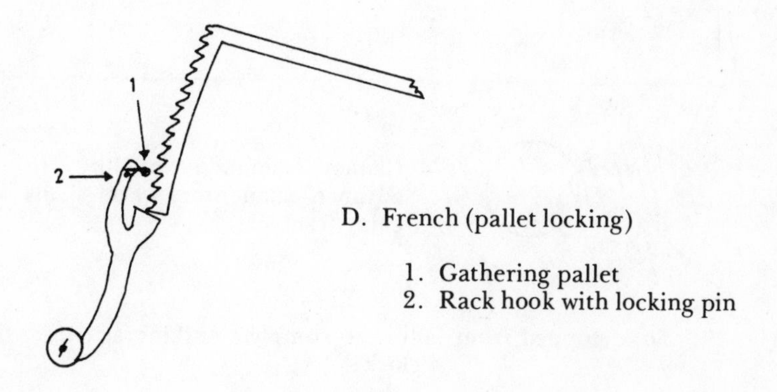


Fig. 19. Locking of rack striking trains (*see also* Fig. 21 and Plates 3, 4 and 5)

the rack falls on to a particular segment of the snail and is gathered up by the pallet and, as one tooth on the rack represents one blow on the bell or gong (and one revolution of the locking wheel on which it is mounted), the appropriate number of blows is struck until the rack is up and the rack hook drops into place with the train locked. If—and here is the distinction from the countwheel system and the great use of the rack system for repeating mechanism—the rack is again released and the snail has not moved significantly, the same

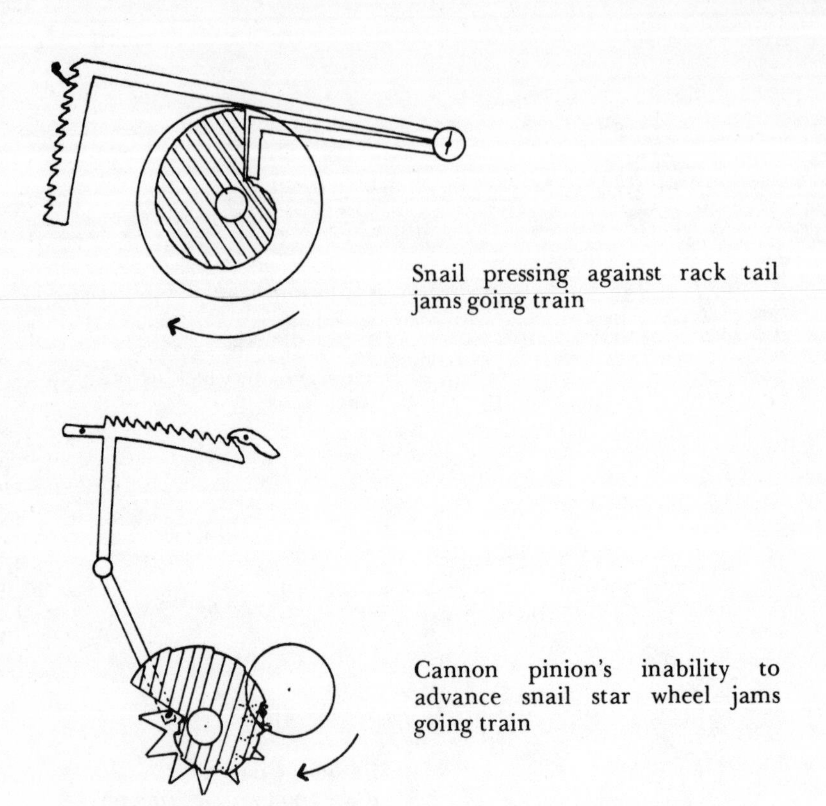


Fig. 20. Clock stopped from failure to complete striking at twelve o'clock

hour will again be struck. If the hands are moved forward several hours without the clock's being 'struck round', the snail will also be moved round with the hands, and the clock will still strike correctly again at the next hour. Conversely, there is an inconvenience; if the rack is released and, for whatever reason, the clock does not strike twelve, the rack is liable to jam against the vertical side of the snail and bring the going train to a halt also. Striking clocks which stop between twelve and one o'clock are always suspect in this direction and the practical test, without dismantling, is to wind the striking train fully and to move the hands back carefully to twelve o'clock. Frequently the clock will then strike and it will be found that the going train also has been released and will

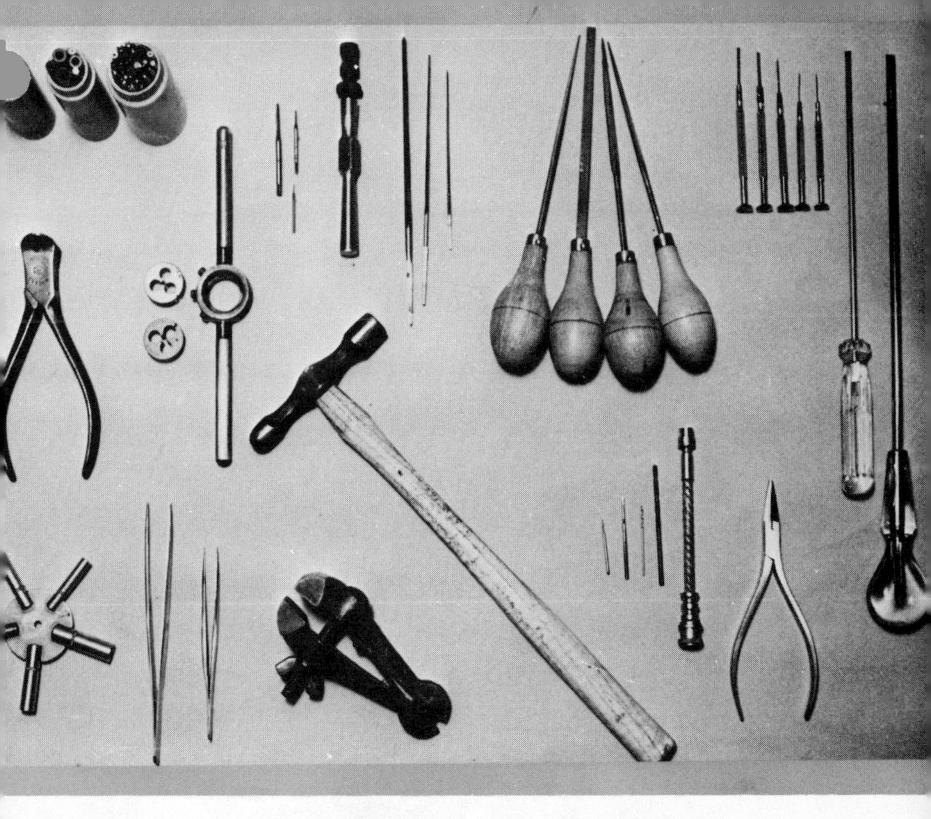

1. Useful Tools
(*top row, left to right*) three sets of punches, taps and tap wrench, three broaches, four files, assorted screwdrivers
(*middle row, left to right*) top-cut nippers, dies and wrench, light hammer
(*bottom row, left to right*) star of keys, tweezers, hand vice, drills and drill stock, pliers

2. French Countwheel Striking (*see* Figs. 16, 17)

3. Old English Rack Striking (*see* Figs. 18, 19, 21)

4. Modern English Rack Striking (*see* Figs. 18, 19, 21)

French Rack Striking with Lifting-Piece. (Movement of a non-repeating carriage clock.) *See* Figs. 18, 19, 21

6. 'Ting-Tang' Striking (English). *See* Fig. 23

Three-Train Rack Chiming (English, flirt release). *See* Fig. 25. The flirt, with large counter-balancing tail, can be seen above the central hour wheel

8. Three-Train Countwheel Chiming (English). *See* Fig. 26

Eureka Electric Clock (dial removed). *See* Fig. 30

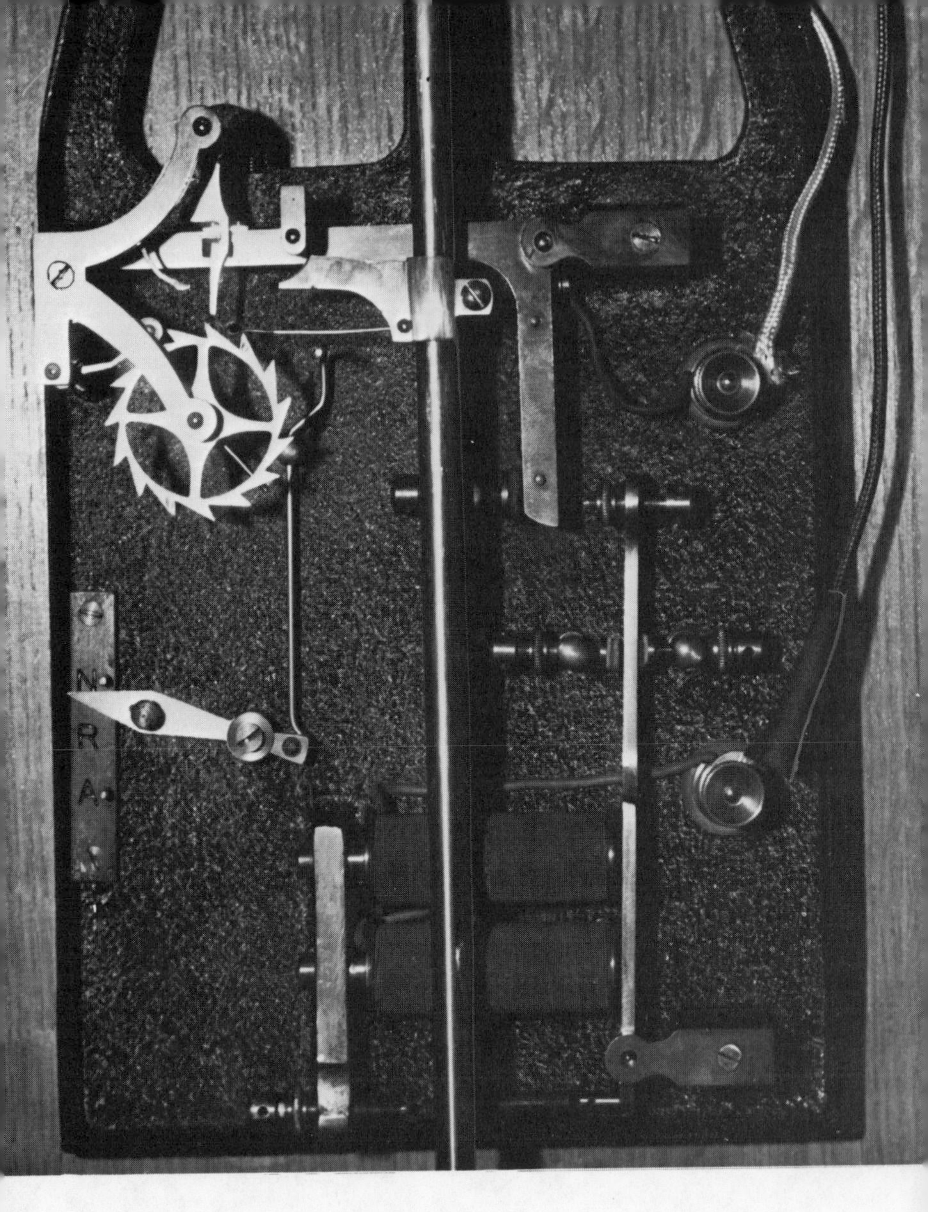


10. Synchronome Master Clock (detail). *See* Fig. 31

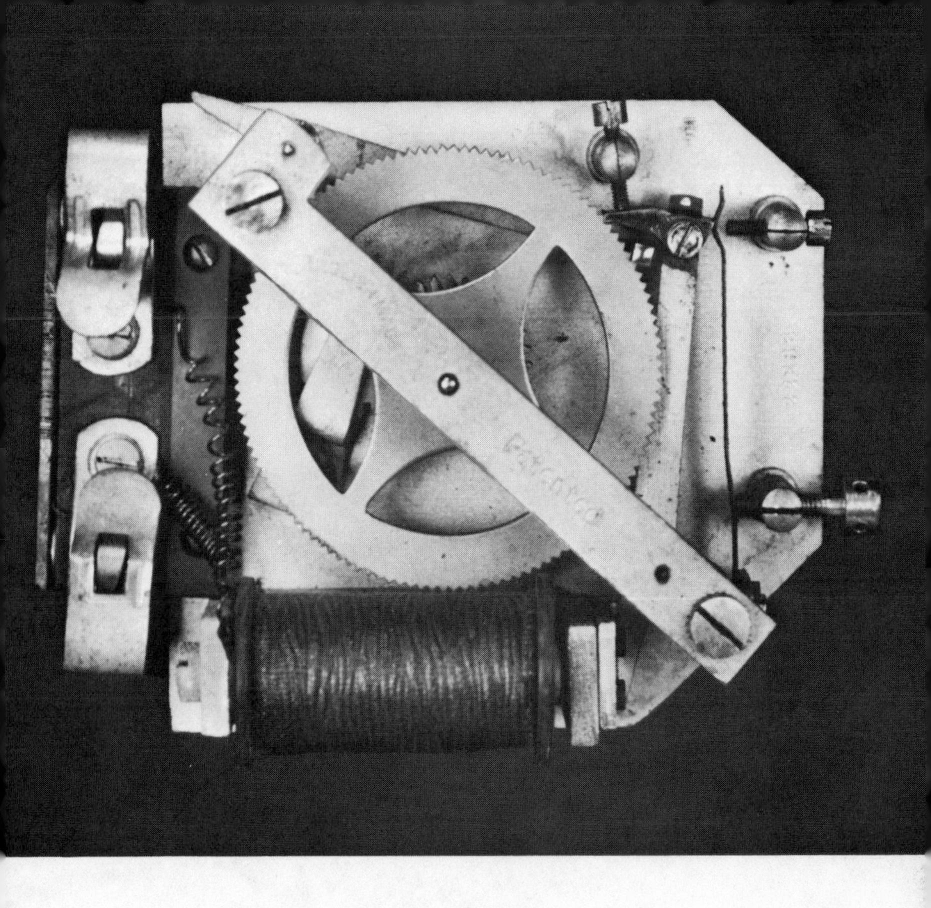

11. Synchronome Slave Movement (from rear). *See* Fig. 32

12. Slow-running Synchronous Motor Movement (*see* Figs. 33, 34). This movement has a buzzer alarm, the springy armature (to the bottom left, above the field coil) being silenced until released by the cam-operated lever above it. Part of rotor and spider are visible at the top of the front plate

13. Magnetically Driven Battery Movement (*see* Fig. 37)

14. Development of a Battery-powered Movement (*see* p. 95). Switching by transistor (the black disc between the two screws to the left of the upper picture) or, earlier, by mechanical contact (lower picture)

15. Battery Movement with Magnetic Balance and Energized Coil (*see* Fig. 38)

16. Movement of a Domestic Quartz Clock. The principal elements shown are quartz crystal in canister (*top right*), trimmer condenser with regulating screw, integrated divider circuit with connecting tags (*centre, partly obscured*), drive motor and gear train to hands. (A digital clock has control and drive circuitry and the display panel array, instead of motor and gear train.)

run—though it will still be necessary to find out why the rack was not gathered at twelve o'clock (often simply because the clock needed winding). This common fault is shown in Fig. 20.

Most rack striking systems are let off by a lifting piece working on pins or cams on the central arbor cannon pinion, just as is the countwheel system, and the lifting piece is joined to the warning piece, which projects through the front plate into the movement and engages with the pin on the warning wheel, *see* Fig. 21. The lifting piece raises the rack hook which, being connected (except in the old English system) with the locking piece, releases the train. The warning wheel runs half a turn until the pin contacts the raised warning piece, and the train cannot be fully released until both lifting piece and warning piece fall. Sometimes, particularly in cuckoo clocks and modern movements (*see* Plate 4), locking is also performed on the warning pin and the wheel carrying the gathering pallet on its arbor is not in fact a locking wheel at all, but this system is easily understood when met with.

Rather different is the letting off mechanism found mainly in French carriage clocks and known as the flirt system (Fig. 22). Here there is no lifting piece. One part of a jointed lever is pulled back against a spring by the lifting pins as they go round and, when released, it flicks back across the face of the movement. Screwed loosely to this flirt is a lever with a stepped catch in the end. The rack hook is extended upwards and has a pin projecting from it, and the catch on the flirt falls over this pin to knock the rack hook sideways, thus letting off the train, as it is released by the lifting pin and flies across. There is never any 'warning' with this arrangement, which is surprisingly reliable and trouble-free.

We have seen how with the countwheel mechanism an extra-wide slot is used in the countwheel to arrange for a single stroke to be struck at the half-hour, and of course an extra lifting pin is used. With the rack system an extra lifting pin is again used, but it is placed nearer the centre of the cannon pinion than is the hour lifting pin. As a result, at the half-hour the rack hook is not moved sufficiently far for the rack to fall; the train is held at warning and released but, as the rack has not fallen, the gathering pallet merely makes one revolution

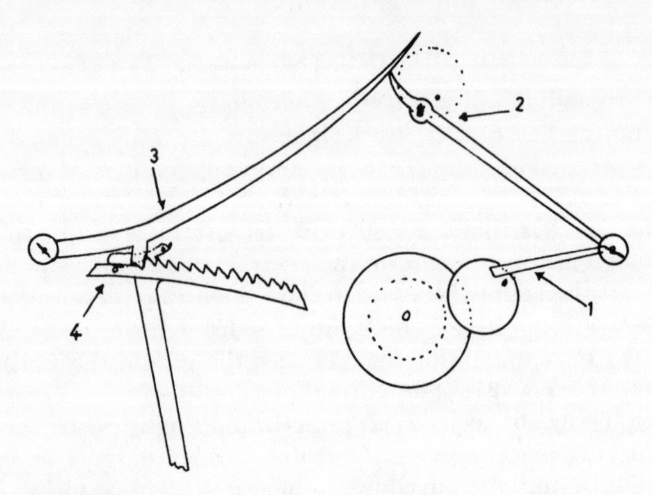


A. Old English

1. Lifting piece working on minute wheel
2. Warning piece (goes through plate) raises rack hook
3. Rack hook when raised lets rack fall, releasing gathering pallet
4. Tail of gathering pallet against rack pin holds train until rack is released

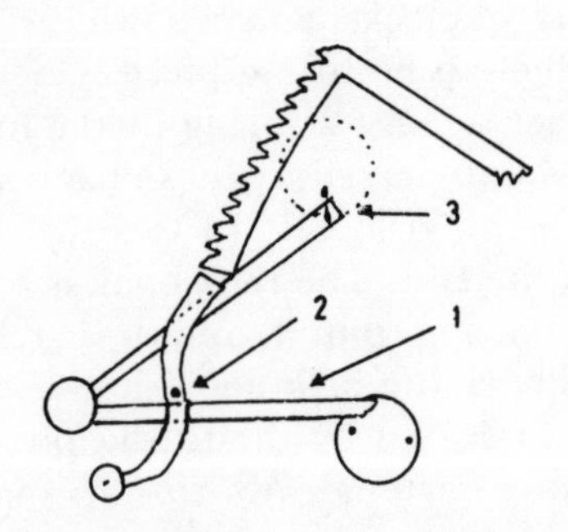


B. French

1. Lifting piece working on cannon pinion raises rack hook by pin and releases train
2. Rack hook and pin
3. Warning piece attached to lifting piece

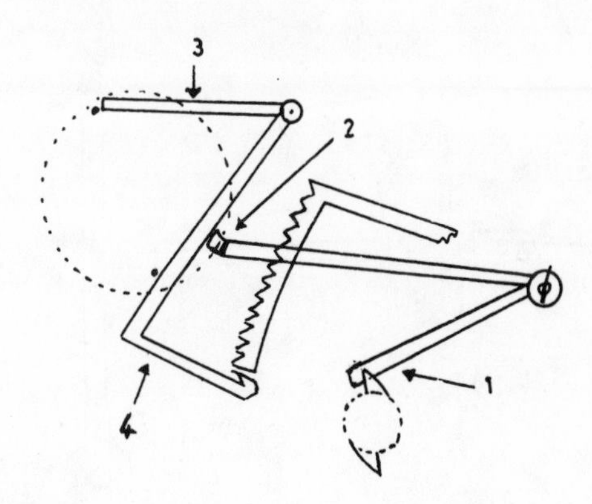


C. Typical Modern

1. Lifting piece raised by cams on centre arbor
2. Warning piece (tip going through plate) raises rack hook
3. Locking piece (through plate) attached to rack hook releases train when rack hook is raised. Locking and warning are on the same wheel

Fig. 21. Some release mechanisms in rack striking (*see also* Fig. 19 and Plates 3, 4 and 5)

and the gong is struck once, before coming to a stop again with the rack still up. Sometimes, however, a short bottom tooth is provided in the rack for the striking of the half-hour. The rack does not, of course, fall down on to the snail.

The half-hour principle is extended in the commonest arrangements for striking quarters from the striking train (rather than full quarter chiming from a separate train). Usually two hammers and bells or gongs are employed, and so the movements are loosely known as 'ting-tang' (*see* Plate 6), although in fact the 'ting-tang' or quarter-striking carriage clock uses a different and more complicated arrangement. Most ting-tang clocks use four graduated lifting pins to set off the train four times in an hour with each quarter being represented by a blow each on a high and a low gong (though more elaborate arrangements are also found). Thus at three-quarters past the hour three pairs of notes are struck and the lifting pin is the next to furthest out on the cannon pinion,

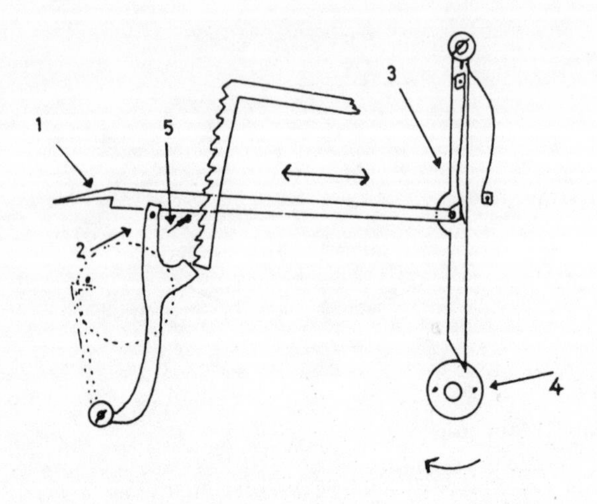


1. Catch in flirt which lodges on rack hook pin and releases rack when the flirt is released by the lifting pin and flies from right to left under influence of its spring
2. Rack hook pin protrudes backwards from an extension of the usual rack hook. Locking piece is on rack hook arbor
3. Flirt sprung where pivoted to lifting piece, so that it will catch the rack hook pin
4. Cannon pinion and lifting pins
5. Extended tooth at back of gathering pallet releases flirt catch from rack hook pin once train is running

Fig. 22. Flirt striking release

releasing the rack for the first three teeth. At the fourth quarter the furthest lifting pin lifts the rack hook clear of the whole rack, which then falls as far as the snail will permit. There is no quarter-striking at the hour. Instead, a lever is moved across, operated by a pin on the minute wheel, to push the high-toned hammer clear of the pinwheel and the hours are consequently struck by themselves on the lower-toned bell or gong. The same part of the rack is used for striking the first three hours as for striking the three quarters. Care is naturally needed to ensure that the high-note hammer is inactivated at the proper time (i.e. just before the hour) and that the minute

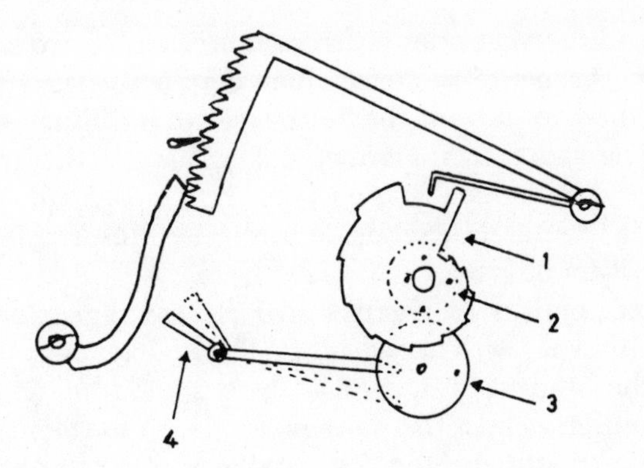


1. Snail mounted on hour wheel (not shown) and with a slot after one o'clock for striking the quarters
2. Cannon pinion with graduated lifting pins governing fall of rack at quarters
3. Minute wheel with pin to operate high-note hammer silencing lever at each hour
4. High-note hammer silencing lever, which presses hammer out of action at the hour (position shown by broken line)

Fig. 23. Outline of 'ting-tang' mechanism (shown striking half past one)—*see* Plate 6

hand is so placed as to correspond to the next quarter lifting pin on the cannon pinion. A special form of snail has to be used for these clocks to provide for the fall of the quarters after one o'clock; it has a notch after a brief raised one o'clock segment (Fig. 23).

Quarter-striking carriage clocks employ a different method, using two racks and snails, so that they can repeat the preceding quarter as well as the hour on pressure of a button which releases the rack hook. The striking of hour and quarter at each quarter (during normal running) is known as 'grande sonnerie', and requires a specially large mainspring and suitable train—such clocks are expensive and relatively rare. Striking in ting-tang fashion (without the hour except at the hour) is known as 'petite sonnerie', but these carriage clocks usually repeat (as opposed to regularly striking) hours as well as quarters. In fact all such clocks are arranged so that they might (with correct trains) strike hours and quarters all the time, but the petite sonnerie clock employs a lever from the minute wheel to prevent the hour rack from falling save at the hour. (The same arrangement is also used with flirt-release clocks striking the half-hours.) In the true grande sonnerie clock, this lever can be set aside if required to permit hours to sound at every quarter.

The mechanism of quarter-striking carriage clocks is of flirt-release type, with no warning. The quarter rack is pivoted behind the hour rack and falls on to a quarter snail on the cannon pinion, whilst the hour rack falls on to the hour snail, mounted on a starwheel in the usual way. At the quarters, the quarter rack is released and the train set off by the flirt action on the double rack hook, but the hour rack (except in grande sonnerie) is prevented from falling by the lever below it—this lever is sprung into position beneath the hour rack and only moved aside just before the hour. At the hour, for hour striking, only one hammer is used, but the other is silenced by a different arrangement from that in the usual ting-tang clock. There are hooks on the ends of the hammer arbors, and these can fall into a slot on a pivoted arm attached to the front plate. This arm is engaged by a pin on the foot of the hour rack so that when the latter is gathered up (or held by its lever), both hammers are free to sound, the high-note hammer hook falling into the slot on the arm, but when hours are being struck the arm falls so that the high-note hammer's hook is obstructed (Fig. 24). There is a double gathering pallet, to operate on both racks, and the rack hook, with its two surfaces, is so shaped that it will only hold the quarter rack up when the hour rack is already held. Thus, if the clock is striking grande sonnerie, the usual sequence is for the racks to fall, coinciding in their first four teeth, but the quarter rack cannot be held until the hour rack is fully gathered. The hour is struck on one gong, the pivoted arm having been released when the hour rack fell, and when the hour rack is fully gathered the pivoted arm is raised so that both gongs can proceed to sound the quarters until the quarter rack is in its turn gathered and held.

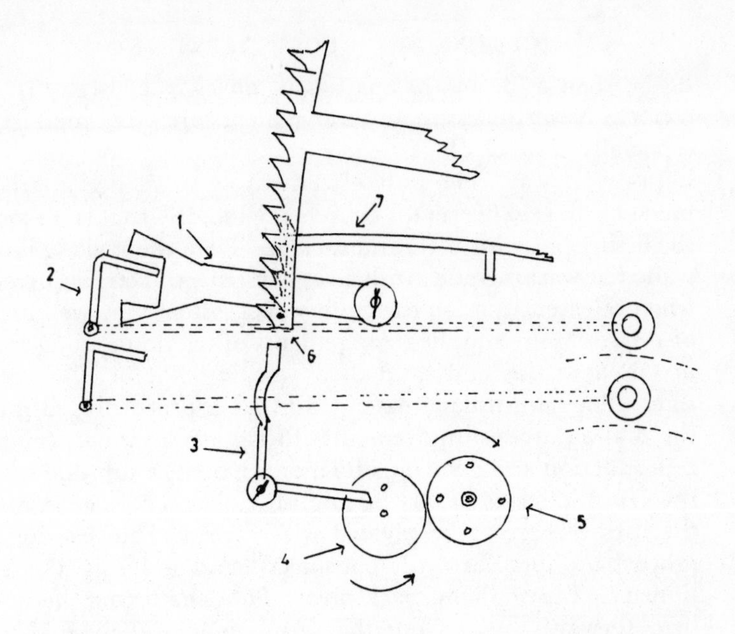


1. Free pivoted lever, raised when hour rack is raised, and slotted to receive low-noted hammer
2. High-noted hammer, free when hour rack is raised, but obstructed by the lever when it falls with the hour rack for hour striking
3. Lever which prevents hour rack from falling save at hours
4. Minute wheel with pin to displace rack-restraining lever at hours so hour rack can fall
5. Cannon pinion, with lifting pins equi-distant from centre (since quarter-striking is controlled by a quarter snail)
6. Pin, at rear of hour rack foot, which holds pivoted lever up when rack is up
7. Quarter rack behind hour rack.

Fig. 24. Silencing of high-noted hammer in quarter-striking carriage clock

*Silencing the Strike*

Reference has already been made to the likelihood of a clock's stopping if the strike is put out of action by being run down or failing to work, as, for example, if the gathering pallet comes loose on its arbor (for these arbors are often not squared). Many movements include a flexible rack tail to guard against this—the springy tail being shaped so that it will be brushed aside by the snail as it moves from twelve to one o'clock. Nevertheless these arrangements are often ineffective, especially if a clock is nearly run down or not going too happily anyway. The only effective way to silence a striking mechanism is to prevent it from being let off—that is, to fasten the lifting piece (or flirt) and rack up. Then the striking is held stalled at warning and, with a rack system, will strike correctly when released, though the countwheel system may well be out of phase. Many simple arrangements of friction-sprung levers are used to this end, and their working will be obvious on sight. They are usually extended to a tip protruding through the dial on modern movements. In older movements (and on reproduction and good-quality modern ones) a sub-dial with a revolving index is usual. In English clocks it was general for the lifting piece to be pivoted with considerable freedom to move back and forward (endshake), and a spring was tensioned against its long back pivot. Thus the lifting piece was held forward, away from the lifting pins and incapable of engagement with the rack hook whose movement released rack and train. A pivoted lever pressed the front of this spring-loaded arbor inwards if it was required for the pieces to engage and the clock to strike. This lever ended in a projecting pin locating in a slot or on a cam to whose arbor was attached the dial index. The flirt release, on the other hand, is usually rendered inoperative by a sloped lever which prevents the loose arm of the flirt from falling far enough to knock the rack hook away from the rack. This lever has in grande sonnerie clocks extensions to control the hour rack restraining piece and also to silence the quarter striking if required, and it normally goes through the base of the movement and indicates against markings on the cover-plate in the bottom of the case.

*Quarter Chiming*

Full chiming with a separate train may employ a countwheel or a rack system. The countwheel (having only four divisions) has been usual in modern times, but the common arrangement in English clocks of the 18th and 19th centuries was to use a rack mechanism. Either way, the sequence of events is that the chiming is let off by a lifting piece at every quarter, and the chime train itself releases the hour train,

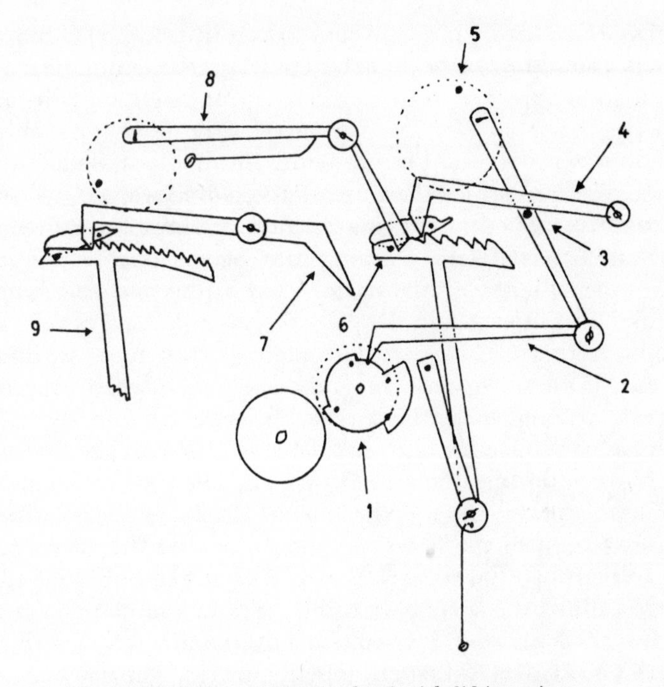


1. Quarter snail on minute wheel with lifting pins
2. Lifting piece (may be replaced by a flirt as in Plate 6)
3. Pin on quarter warning piece which raises quarter rack hook
4. Quarter rack hook
5. Quarter warning wheel
6. Pin on quarter rack which operates hour warning piece. (To rear is pin locking gathering pallet)
7. Hour rack hook extended so as to be displaced by fall of quarter rack at the hour
8. Hour warning piece
9. Hour rack. (*See also* Fig. 18)

Fig. 25. English rack quarter-chiming. (*See* Plate 7)

holding it at 'warning' whilst the fourth quarter is struck, and letting it run once chiming is completed. There is thus no lifting piece, in the usual sense, for the hour train.

In the old English arrangement (*see* Fig. 25 and Plate 7) there is a quarter snail on the arbor of the minute wheel (which is merely geared 1:1 with the cannon pinion and also bears the four lifting pins, revolving once in an hour). Because there is a snail to control quarter chiming, the lifting pins are placed round the circumference, not proportionately distant from the arbor as they are in quarter-striking devices which have no separate snail. They raise the lifting piece, and so also the warning piece into the path of the chime warning wheel, whilst at the same time the quarter rack is allowed to fall on to the snail. Alternatively, chiming may be released by a small flirt, a sprung lever pushed back by the lifting pins and, when free, flying up to hit and displace the rack hook. There is no warning in this arrangement (*see* Plate 7). When the warning is released, the train runs, the gathering pallet (which, as with old rack striking, locks the train with its tail against the rack pin) revolving once for each sequence of chimes. The quarter rack has a pin projecting forward which, when the rack falls to its fullest extent at the fourth quarter, displaces a lever which is an extension of the hour rack hook, so that the hour rack falls. When the quarter rack is released (at each quarter) this pin also allows the sprung warning piece of the hour train to rise, but the hour train is of course only actually released when its rack also is displaced at the fourth quarter. The mechanism is virtually foolproof once correctly set up, but of course it must be ensured that the hour train is reliably released by the quarter rack, and the minute hand must be fitted initially to correspond with the position of the minute wheel, or incorrect quarters will be struck.

Clocks with this arrangement are reliable but massive and, as with all rack systems, there can be trouble arising from wear in the rack and slip between rack and rack tail—these are nearly always connected friction-tight so that the fall of the rack tail can be set to coincide with the proper gathering of teeth in the rack by the gathering pallet. More recent systems generally employ countwheel chiming with rack striking of hours (Fig. 26). The countwheel is as a rule mounted on the front plate (*see* Plate 8). Arrangements for letting off and for locking vary considerably, but usually letting off is by a long lifting piece acting on the four pins or cam-points of the cannon pinion (not of the minute wheel as in rack chiming). Often a short right-angled lifting piece is employed to move this lever, and it can be safely bypassed if the hands should be

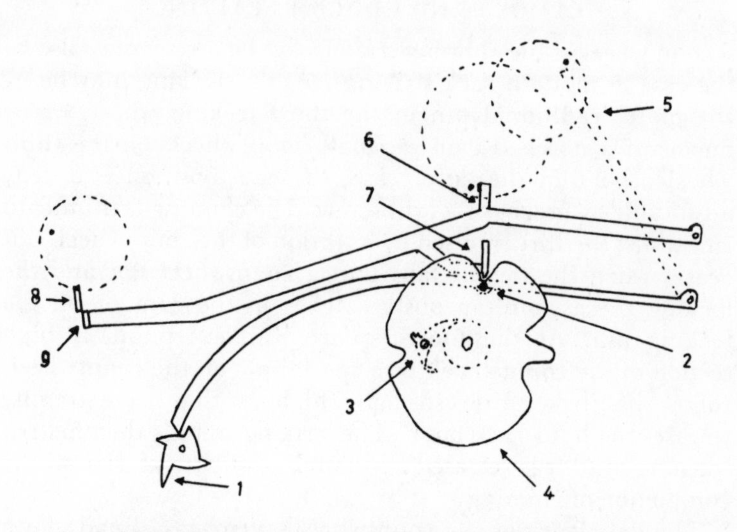


1. Lifting cams on centre arbor, the hour tooth long so as to ensure release of self-correcting catch
2. Pin on lifting piece which engages with self-correcting catch
3. Self-correcting catch on rear of countwheel
4. High raised hour section of countwheel lifts strike flirt sufficiently to release hour rack hook
5. Chime warning piece on lifting piece arbor
6. Chime locking piece protrudes behind from strike flirt
7. Countwheel detent protrudes in front from strike flirt
8. Hour warning piece is an extension of strike flirt
9. Hour rack hook release made by protruding tip of strike flirt

Fig. 26. Type of modern countwheel chiming mechanism (*see* Plate 8)

moved backwards. The lifting piece is connected to or solid with the warning piece, and it also raises the countwheel detent and chime locking piece. These are projections of, or on the arbor of, a long lever going across the front plate towards the striking train and known as the strike flirt. The strike flirt releases the hour warning, and also the hour rack on the hour. Sometimes the strike flirt is fitted with both chime warning piece and locking piece acting on the same pin in the second wheel of the chiming train (as we have seen may also be the case in modern rack striking). Again, locking may be by the more traditional pin in the third locking wheel, or by means of a notched cam, a small hoop wheel, on the third wheel arbor—in the latter case, the cam position is usually adjustable by means of a grubscrew. The chiming continues to run whilst the flirt is raised by a section of the countwheel, but ceases when the detent falls into a countwheel slot and the locking piece simultaneously catches in the cam or on the locking pin. At the fourth quarter an exceptionally high section of the countwheel (or a special pin on the countwheel) raises the strike flirt especially high so that its extremity releases the hour rack hook. The striking train is then held at warning until it is released by the fall of the strike flirt on the completion of chiming.

The deficiency of the countwheel system has already been noted, and arises from the fact that the regulation is in the train itself, so that the striking or chiming must run in sequence, regardless of the position of the hands. A self-correcting device is therefore usually incorporated in count-wheel chiming, and it takes the form of ensuring that when the chimes reach the fourth quarter they are held there until the minute hand also points to the hour, at which an abnormally high lifting pin is used. These arrangements vary. One of the commonest is a small sprung catch (Fig. 26) attached to the rear of the countwheel at the third quarter slot—this catch is caught by a pin on the countwheel detent, so that the count-wheel and train cannot revolve, and it is only released when the detent and flirt are raised more than usually high by the special lifting pin. Alternatively, there may be a second locking piece acting in a second hoop wheel or on the same locking pin—this locking piece is freely pivoted and the other end rests on a cam with a single notch attached to the count-wheel, so placed that the foot of the special locking piece drops into the notch and the train is additionally locked after the third quarter, being again released only by a specially high lift (which raises an extension of this locking piece). From a description, these devices seem more complicated than in fact they are, but it is true that they can cause difficulties in setting up a chiming clock if one is unfamiliar with them. The essential fact to bear in mind is that they depend on an additional locking facility brought in at the third quarter and only able to be released by the fourth quarter lifting. The countwheel and locking or hoop wheels here, as in hour striking, must be correctly placed so that the locking of the train coincides with a slot in the countwheel, and on modern clocks these wheels are usually mounted with grubscrews so that their positions can be adjusted after assembly.

The point has already been made of striking trains generally that the gearing must be precisely calculated so that one turn of the locking wheel corresponds to a stroke on the bell or gong and that, in the countwheel system, a turn of the countwheel corresponds to 78 or 90 strokes (according to whether half-hours are struck) and revolutions of the locking or hoop wheel. Chiming trains follow similar stipulations, save that many chiming blows are equivalent to the single stroke of the striking system. These blows are not haphazard and irregular, however. Whatever the tune chimed, it has a cumulation of sequences from the first to the fourth quarter, ten such sequences being required over a full hour—one at the quarter, two at the half, three at three-quarters (making six in all) and four at the last quarter. The locking wheel has therefore to revolve ten times for each hourly revolution of the chiming countwheel. The chimes are sounded by some form of pin-barrel which, on modern clocks, is usually mounted, with the connecting gears, outside the movement's plates. These connecting ratio wheels engage with the countwheel arbor. The barrel does not set out the complete order of chiming, any more than the pinwheel of a striking clock is provided with 78 or 90 pins. This would be unnecessarily cumbersome and expensive in materials. Instead, it sets out the order of five sequences (up to and including the second sequence of the third quarter), which are exactly repeated over the rest of the hour, and thus it is geared 2:1 with the countwheel arbor. In setting up, a ratio wheel is temporarily removed and the barrel turned until either the first quarter or the third sequence of the third quarter (in practice identical) is about to be struck, according to the position of the countwheel (*see* Chapter 8).

Finally, it should be said that synchronous electric striking and chiming clocks, now not often met with, work on similar lines, power being provided for the sounding trains either by small springs kept wound by the motor, or by idler wheels automatically thrown into connection between the going and sounding trains when required.

*Alarm Mechanisms*

If the striking clock is older than the visual indicator, it is itself most probably a development of the alarm, the oldest of all. The simplest alarm imaginable is a bell struck mechanically but let off manually at a predetermined time. Such a device is then attached to a continuously running mechanism which sets it off automatically. The alarm train consists of a weight- or spring-driven barrel and usually one intermediate wheel to give it a reasonable run when let off. The final wheel resembles a recoil scapewheel and is engaged with a pair of pallets, often a shaped block working on one wheel tooth only, to whose arbor is directly attached the alarm bell hammer. Thus essentially we have an over-powered escapement which flutters violently, being very little under the control of the hammer, which may be seen as a rudimentary pendulum. There is, of course, no crutch, and the hammer pivots and bearings are subject to the same extensive wear as were the old knife-edge and crutchless pivoted pendulums attached directly to pallet arbors.

The alarm power is generally derived from a separate spring or weight, but a common modern practice is to use one spring to drive both going and alarm work. In this case the barrel top is replaced by a gear wheel, usually connected into the alarm train, which is fixed to the barrel arbor but linked by click work to the train (Fig. 27). Simple stop-work is provided so that this gearwheel can only revolve once when the alarm goes off; thus, even if the alarm is not manually stopped, its running time is limited so that there remains power in the spring to drive the going train, and the spring will have to be rewound before the alarm can go off again. This stop-work is of 'one-way' type in that it does not of course prevent winding of the mainspring for the six or so turns necessary to secure full power.

Electric alarms feature a buzzer which sounds either when the current is switched to it by the alarm-setting device of the

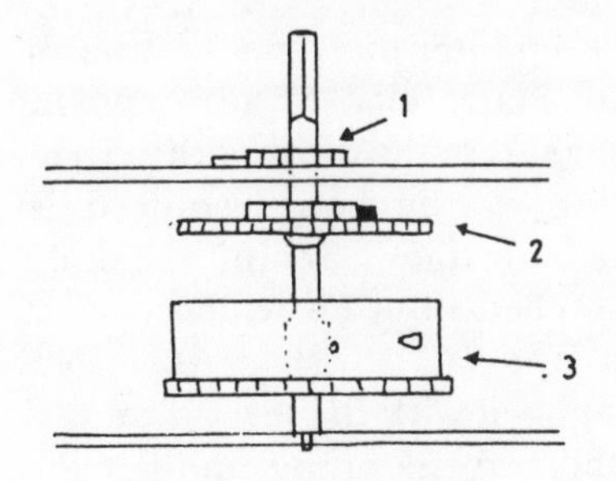


1. Click work and simple stop work allowing alarm to unwind only one revolution
2. Alarm greatwheel connected to arbor by click work
3. Going barrel driving going train

Fig. 27. Going and alarm trains powered by single spring

clock, or a buzzer permanently vibrating against the field core of the synchronous motor, but prevented from sounding, until required, by a lever interposed by a twenty-four or twelve-hour cam, whose position is varied according to the setting, the arrangement of which is conventional. Battery alarms may employ a transistor circuit producing oscillations in an iron-cored coil, whose fluctuating magnetism causes a springy armature to vibrate and sound.

Clocks from the 19th century onwards have generally been provided with a silencing lever or button. This takes the form of a lever obstructing either the hammer or the alarm scape-wheel. It is often so shaped and sprung as to be released automatically when the alarm is wound. Some modern clocks feature an intermittent or 'repeater' alarm (also known familiarly as a 'snooze'), which sounds every two or three minutes for a period after being let off, striving to jolt the dozer into life or a nervous breakdown. The mechanism consists of an idling pawl and tail working on a ratchet fixed usually to the third wheel of the going train. The pawl tail obstructs the hammer, temporarily silencing the alarm, whenever the pawl is on a high tooth of the ratchet; as the ratchet is turned by the going train the pawl falls back and the hammer is released, allowing the alarm to sound.

There are two related setting mechanisms. The first one, whose use is indicated by a friction-tight circle of clockwise numbers in the centre of the dial, is obsolete, but it was employed, though not universally, until at least the end of the

19th century and may be found even later on some 'postman's alarm' clocks (weight-driven wall alarm clocks with exposed pendulums and weights). The alarm is set by moving the circle so that the figure corresponding to the hour required is immediately below the tail of the hour hand. Behind the dial, solid with the ring, is a thick notched disc, on which rests a pivoted lever connecting with the alarm scapewheel or hammer. The notch when it is engaged with the lever is on the opposite side of the disc from the corresponding number on the circle. Thus if the lever's (fixed) position is at one o'clock, the notch will be behind the number 7 on the ring on front of the dial. The disc and front ring rotate with the hour hand. When the end of the lever falls into the notch, the other end of the lever swings and releases the alarm train. The notch always permits sounding at the same position but, according to the time for which it is set and the time shown by the clock at the moment of setting, it takes a longer or shorter time to get there. The device is not easy to explain, but can be clarified by inspecting an actual example.

The later universal device also employs a collet with a notch, sloped on one side. This disc is on a sprung arbor so that it can be set as required. Free on the same arbor, but pressed against the collet by a spring, is a wheel with a steel nib or projection. When the notch and this projection coincide, the wheel jumps up and this movement brings it free of the lever sprung to obstruct the alarm so long as the wheel presses on it. The lever therefore moves clear, and the alarm goes off. In the normal course of running, the alarm wheel is pressed down against the slope of the collet's notch so that the alarm is in due course silenced and ready to be let off again after re-winding. In the modern cheap alarm clock synthetic materials may be used for collet, alarm wheel or both. The alarm wheel may be the central hour wheel or a wheel geared to it—hence it is sometimes observable that the hour hand moves in and out each day. The spring pressing the alarm wheel and collet together may also do duty as the interrupting lever, being a strip of steel passing over the front plate, pressing on the alarm wheel and then being bent over into a hook to obstruct the hammer. The roles of collet and alarm wheel are often reversed; a synthetic hour wheel having a recess cut into it jumps up against a pin fixed into the setting arbor, the wheel rather than the arbor thus bearing the collet, and the arbor rather than the wheel having the projection. It should be added that the jumping of the alarm wheel may also be used to make a contact and complete the circuit to an electric bell or buzzer. If the principle is understood, such variations are clear enough in practice.

The alarms found in carriage clocks—many clocks were designed so that alarms could be inserted as an option at manufacture—have certain distinctive features (Fig. 28). As a rule, the alarm wheel is taken from the hour wheel by an intermediate wheel. The ratio is normally unity, the alarm wheel also revolving once in twelve hours, and the purpose is to move the alarm down so that its subsidiary dial can be at the bottom of the face. The alarm wheel, mounted on the set-alarm arbor with the notched collet, is held by a sprung rocking arm pressing up from below, until the wheel's projection jumps into the collet notch and this end of the arm is freed. The other end of the arm has a pointed adjustable screw impeding a long trigger descending from the alarm hammer and pallet arbor, and this screw is used to arrange for positive locking of the hammer and prompt release at the appropriate moments. Movement of the hammer is controlled by two leaf-springs, acting on pins each side of the hammer trigger, near the arbor. If these springs are unsuitably tensioned either the hammer will jangle on the bell or its pallets will jam on the alarm scapewheel.

Finally, it will have been noted that these letting-off devices all employ a collet with a notch of which one side is vertical and the other is sloped. It is for this reason that alarms must only be set in one direction, for if the collet is turned so that the vertical face of the notch butts on to the projection of the alarm wheel (which is directly connected to the going train), damage is liable to be done to both alarm and going mechanisms as the train is forced violently backwards. The better class of alarm, particularly in the case of carriage clocks, provides the set-alarm square with a stout ratchet and click so that it is virtually impossible for the alarm hand to be turned in the wrong direction. The majority, alas, merely provide an indicating arrow which is all too often ignored.

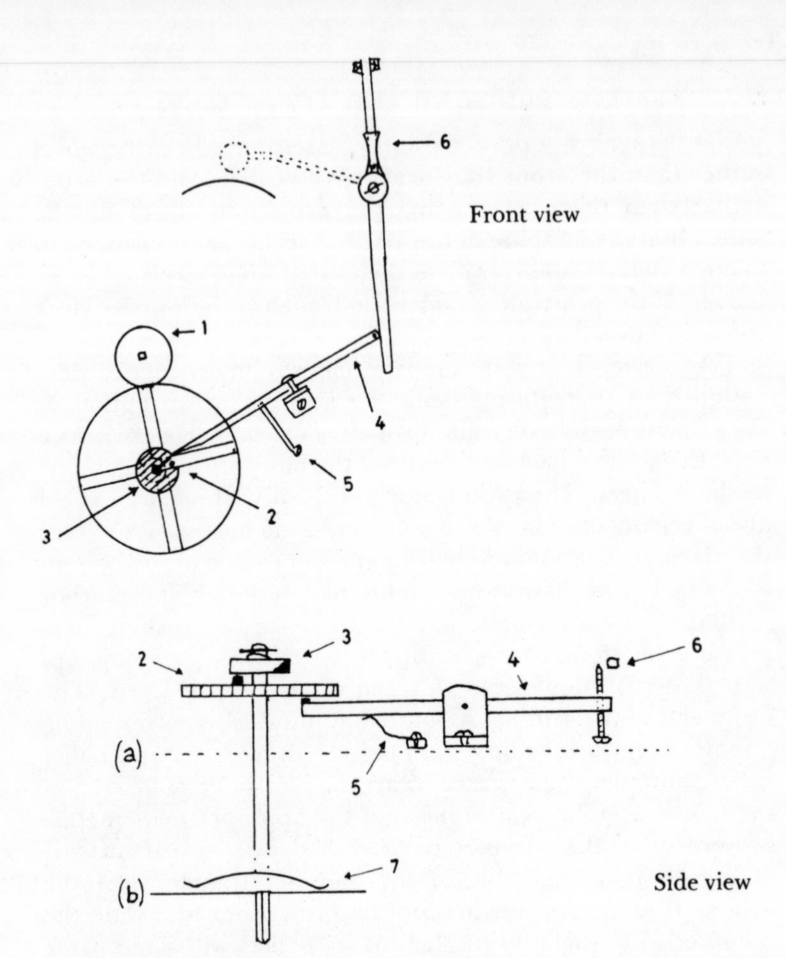


1. Transmission wheel from hour wheel
2. Alarm wheel with projection to press up into alarm cam notch
3. Notched alarm collet fixed to arbor, which is the set alarm square
4. Rocking arm sprung to press alarm wheel against cam, and with screw to engage end of hammer trigger when alarm is off
5. Rocking arm spring
6. Hammer trigger and springs to limit its movement
7. Alarm tension spring, keeps alarm's setting whilst alarm wheel is turned by motion wheels

a. Front plate of movement
b. Back plate of movement, with alarm set-square

Fig. 28. Carriage clock alarm mechanism

*Chapter Five*

# ELECTRIC CLOCKS

## *Types of Electric Clock*

Electric clocks have been in use, at least on a small and experimental scale, for well over a century, and many times it has seemed to inventors and pioneers that they were poised to supersede spring and weight-driven clocks, surpassing them in accuracy, dispensing with the chore of winding, and requiring less maintenance. Yet it is only very recently that electric and 'electronic' clocks have made a really sizeable dent in the clock market. There are several reasons for this, the principal one being that, whilst the most accurate clocks known have been electrically powered or almost entirely electronic clocks, it is only within the last few years that the micro-electronics industry has been able to produce versions of them at a size and a price to suit the home. A further factor has been the development of small, long-lasting electric cells which could free clocks from dependence on massive large-capacity dry batteries or accumulators, or on the domestic mains supply. The production of watches has of course been affected similarly. The mass-produced electric article will now give a better performance than its mechanical equivalent of comparable price. For the traditionalist, who likes to see a swinging or revolving balance or pendulum, many conventional models have meanwhile been electrified to give the best of both worlds.

It is not possible here to outline the evolution of electric clocks or even the very wide variety now available. All that can be done is to mention one or two popular clocks of the past and some main types now in use. For practical purposes, it remains to be seen how far electronics will come to dominate the electric clock industry and how far the repairer of electric or electronic clocks will be the high street electrician, or indeed the specialist manufacturer to whom many clocks are even now returned for repair or, often, replacement. To an unprecedented extent, electric clocks, though often retaining simple mechanical devices for actually showing the time, are now composed of assembled units or modules. Repairing, or even distinguishing, a defective single part in such units is simply not economic for much of the time and it is in the customer's interest for the unit itself to be replaced through the agency of a retailer. Electric clocks are more and more part of the 'disposable economy'. The economics of the industry are increasingly such that it becomes cheaper to replace a whole movement (i.e. everything but case and indicators) than to repair it, though there are areas where the amateur who is prepared to discount overheads can take a practical interest and achieve some success.

Electric clocks fall into certain fairly well-defined groups. There is the 'motor' driven direct from the mains or, in portable form, by a tuning fork kept in vibration by a transistor and magnet. There is the mechanical movement with a spring or weight which is regularly rewound by a magnet or motor. There are the direct contact magnetic impulse types, now largely superseded, and the gravity impulse type in which a falling arm gives the impulse but is electro-magnetically reinstated. Then, at the present time, there is the magnetic balance and, of course, the more purely electronic (i.e. without moving parts) quartz movement which has only recently entered the domestic market at a popular price-level. Some of the older types involving direct electrical contact between metal points have also been updated so that this switching is performed by the effect of a magnet as it approaches contacts in a small vacuum tube (a 'reed switch'), or by the 'relay' effect of a transistor which will detect the minute current induced in a coil by a magnetic balance or pendulum bob and initiate an impulse of increased current. The great majority of electric clocks follow a plan opposite to that of ordinary clockwork, in that they use a fast oscillator actually to drive a reducing gear train which ends with the hands or other indication (*see* Fig. 6, p. 28). Some employ light-emitting diodes or liquid crystal displays to give a digital read-out, and thus reduce or dispense with mechanically moving parts altogether.

*Synchronome, Eureka and Bulle Clocks*

The Synchronome type of clock is not superseded as an

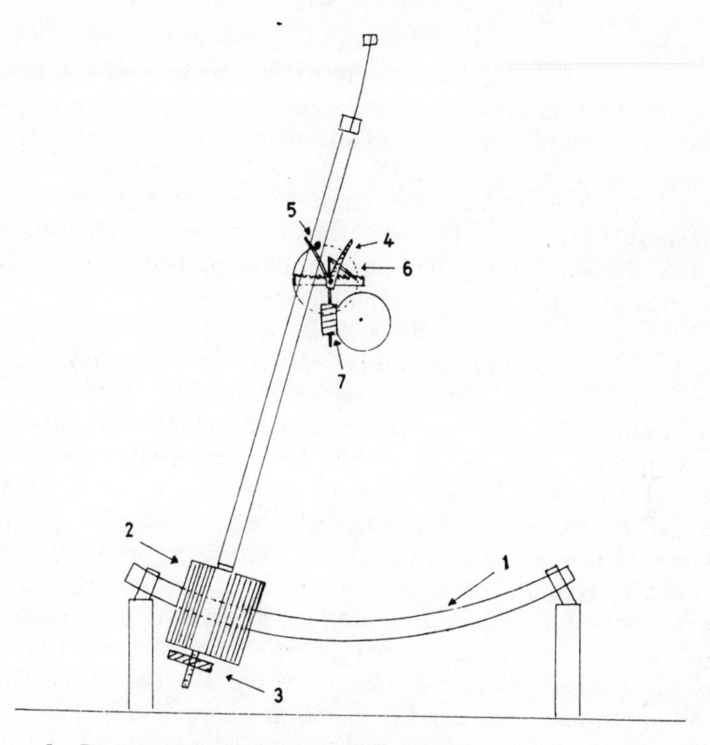


1. Permanent magnet fixed to clock base
2. Coil, acting as pendulum bob, passing over permanent magnet
3. Heavy regulating nut
4. Forked contact, one half insulated
5. Pin contact in pendulum rod engages with forked contact
6. Pawl, raised from contact arbor, drives crown ratchet
7. Worm by which ratchet revolution drives train

Fig. 29. Basic Bulle mechanism

indicator of time on dials dispersed round premises, but we will consider it first with the older Bulle and Eureka models as being nearest to clockwork as we have so far seen it. These clocks possess pendulums and balances which mechanically drive wheels, though the Synchronome count wheel is not in fact part of a gear train but merely governs the intervals at which impulse is given to the pendulum. The drive takes the form of a hinged or pivoted pawl worked by the pendulum and turning a ratchet. In the Eureka clock (*see* Plate 9) one end of a pivoted pawl rests on an eccentric cam, part of the large balance wheel arbor, and so is moved in and out as the balance swings. Any such arrangement involves friction and where (as in the Bulle and the Eureka) the ratchet is part of a train, there is considerable interference with the oscillator's natural vibration. The Synchronome device of an isolated wheel driven by a jewelled pawl and not part of a gear train, was designed to minimize this interference.

The sole use of electricity in Bulle (Fig. 29) and Eureka (Fig. 30) clocks is to give impulse to pendulum or balance at each oscillation. This is accomplished by making the pendulum bob or two balance arms the poles of an electro-magnet, or a coil which swings over a permanent magnet. Either way, when the circuit is switched on near the centre of the vibration, the pendulum or balance is attracted to a static piece of iron or a magnet, thus receiving an impulse. The switching is so arranged that the current ceases as magnet and armature come opposite each other, so that the pendulum or balance can swing past as the magnetic field collapses. In both clocks a direct contact is used. In the Bulle it is between an insulated pin in the pendulum rod and the metallic half of a V-shaped pivoted contact block on the same arbor as the pawl which operates the crown (ratchet with raised teeth) wheel. In the Eureka clock the contacts are a vertical leaf spring on the base of the clock and a special contact pin on the balance connected to the coils round the balance wheel arms. One half of this pin is insulated so that the balance makes no contact on its return journey.

Both these clocks require large-capacity dry cells which have to be connected with the correct polarity—reversing the connections does no immediate damage, but the Bulle clock will not work. Suitable batteries for these and the Synchronome can now be difficult to obtain, but a rectified low tension supply from the mains (such, for example, as a cheap mains unit from an electronic calculator) will serve instead provided that the contacts are adjusted so that the arc of pendulum or balance does not become excessive (i.e. violent, or sufficient to move two teeth of the ratchet at a time, instead of one tooth). Both clocks are regulated conventionally, the

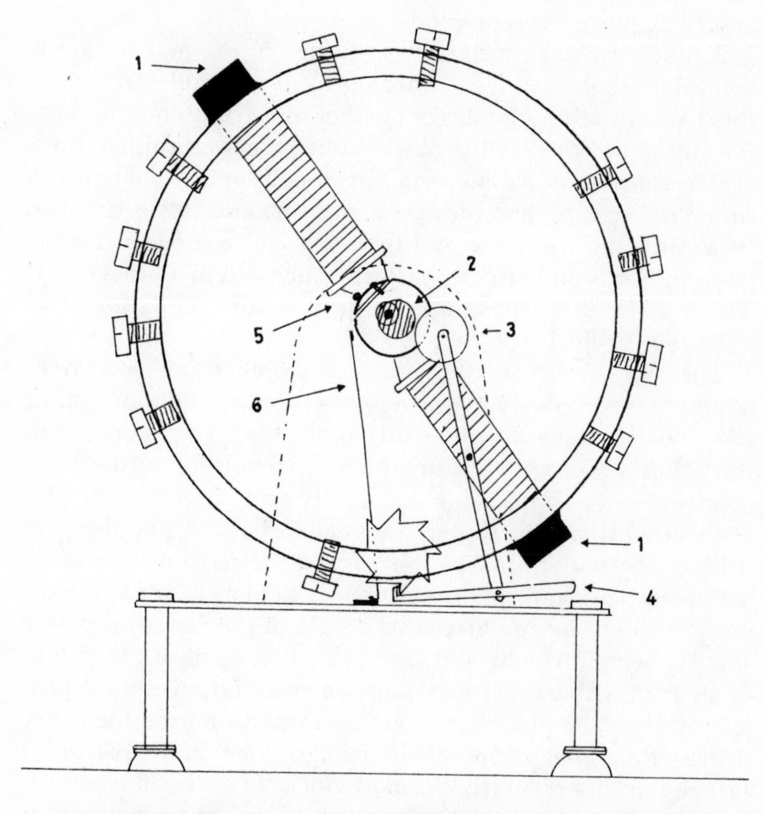


1. Poles of electro-magnet which forms arms of balance wheel
2. Eccentric cam at front end of balance staff
3. Roller on pivoted arm which, moving on the eccentric cam, operates ratchet pawl driving the train
4. Pawl
5. Silver pin contact (connected to end of coils) insulated from balance staff
6. Spring contact mounted on insulated block

Fig. 30. Eureka clock mechanism (*see* Plate 9)

Bulle by a large nut on the pendulum rod, and the Eureka by an index on the hairspring, operated by a rack and pinion. In both, the metal frame of the clock forms one part of the electrical circuit and it is essential that any insulating washers are left in place or made good.

A modern clock similar to the Bulle has the magnet as the pendulum bob, swinging through a static coil. Originally there was switching by direct contact of a fixed point and the moving tail of the driving pawl. In later models this system is replaced by an electronic switch; two coils are used, in one of which the passing bob induces a current, and this is detected by a switching transistor and the other coil is used to give the impulse. Coils and circuitry are all concealed in a brass cover. This is an interesting example of the electronic updating of an older type of clock.

The Synchronome (Fig. 31) and allied systems were triumphs of simplicity, using very heavy long pendulums (most often beating seconds) whose mechanical function is limited to the effort needed to sustain their own motion, turning the countwheel and unlatching the gravity arm. Correspondingly, the electrical side of things is extended beyond impulsing to driving the dial trains by means of an electro-magnetically activated pawl and ratchet of precise design. The impulse to the pendulum is unrelated to the state of the electrical power supply, being given by a gravity arm. It consists of the falling of an arm, with a free roller on it, down an impulse plane projecting from the pendulum rod and retreating from the roller just as the roller drops on to its face (*see* Plate 10). This impulse occurs regularly (in most clocks at every 30 seconds), but at each revolution of the countwheel, not at each vibration of the pendulum. The force needed to keep the Synchronome pendulum (whose bob may weigh 16 lb) in motion through its small arc is extremely small, for very little energy is wasted. These are amongst the most accurate of all commercial pendulum clocks.

A difficulty with direct contacts is the force required to make them effective over a period. This interferes with free oscillation in the Bulle or Eureka systems, and the contacts eventually oxidize from sparking and give trouble. The Synchronome system is more reliable because it involves less

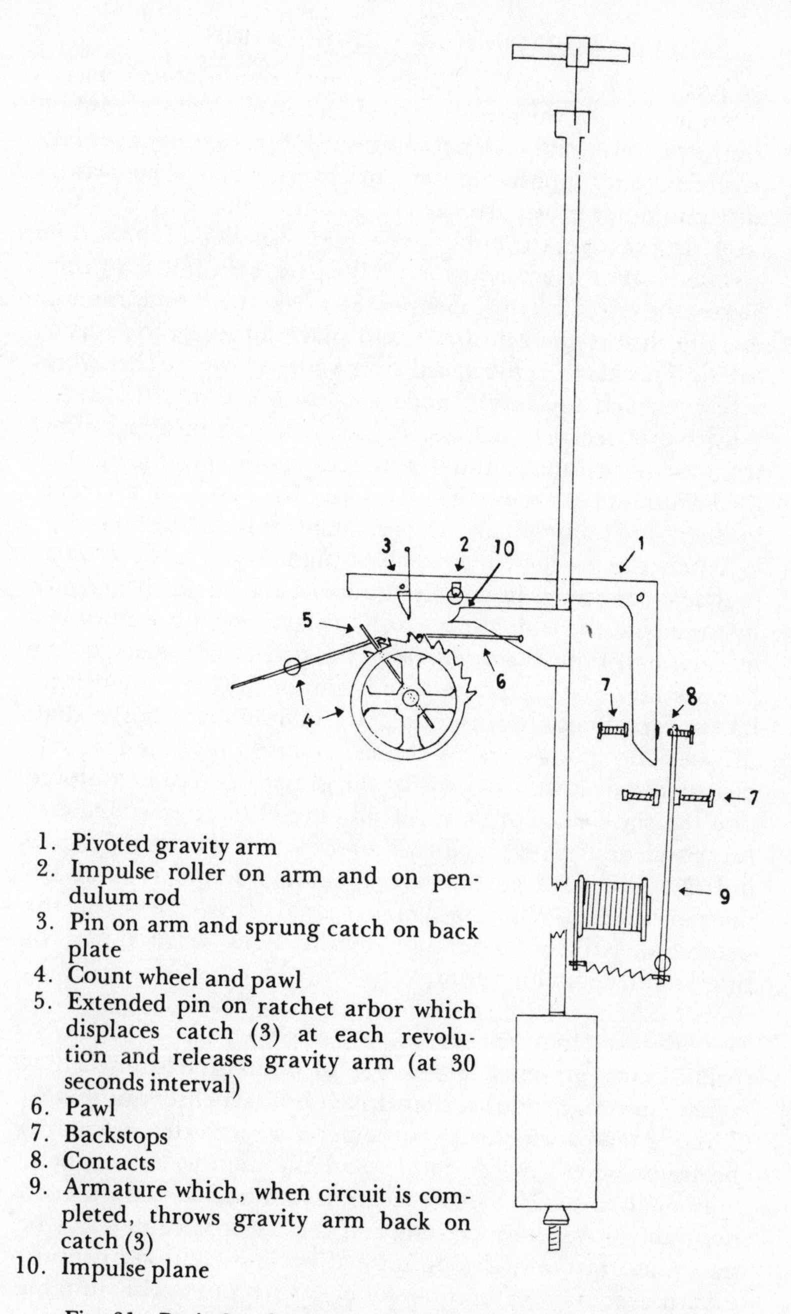


1. Pivoted gravity arm
2. Impulse roller on arm and on pendulum rod
3. Pin on arm and sprung catch on back plate
4. Count wheel and pawl
5. Extended pin on ratchet arbor which displaces catch (3) at each revolution and releases gravity arm (at 30 seconds interval)
6. Pawl
7. Backstops
8. Contacts
9. Armature which, when circuit is completed, throws gravity arm back on catch (3)
10. Impulse plane

Fig. 31. Basic Synchronome-type mechanism (*see* Plate 10)

frequent contact and because it has the contact closure pressure given by the gravity arm rather than by the pendulum itself; the lower end (or another point) of the gravity arm falls firmly on to its opposite contact and stops, whereas such resistance could never be permitted in a contact made by the oscillator. In practice, the opposite contact is the top of the sprung armature of an electromagnet. When the arm falls and completes the circuit, this magnet attracts the armature and thereby throws the arm back into place on its lightly sprung catch. This catch is displaced by a vane on the ratchet wheel arbor at each revolution and is so sprung that it holds the gravity arm securely but can be released with minimal effort from the pendulum. Similarly, to reduce friction, the pawl on the pendulum is set so that it lightly brushes over a ratchet tooth before pulling it and thus encounters little resistance.

When the electrical contact is made, any other electromagnets connected in the circuit are also energized, and it is by these that the trains and hands of slave clocks are advanced at each half-minute (or whatever the period of the clock may be), *see* Fig. 32. The power supply can be dry cells or accumulators, increasing in proportion to the number of 'slave' dials driven. The voltage has, of course, no effect on time-keeping, provided that it is sufficient for the gravity arm to be restored and the slave dials to move. There are various arrangements for mains units with stand-by batteries, and warning lights to indicate the state of the power supply. Regulation of the master clock is by the conventional screw beneath the pendulum bob, but a timing tray (for light weights) may be fitted on the pendulum rod.

### *Synchronous Motor Clocks*

The other principal older type of domestic clock (still, of course, in production) is that driven by a synchronous motor. This is a motor which can run only at a speed determined by the frequency of the *ac* mains and the number of magnetic poles built into it. The frequency standard is scrupulously maintained over long periods and these clocks cannot but be highly accurate in the long term if no interruption of supply takes place. As they are merely electric motors with suitable reduction gear trains they are, however, of limited interest to

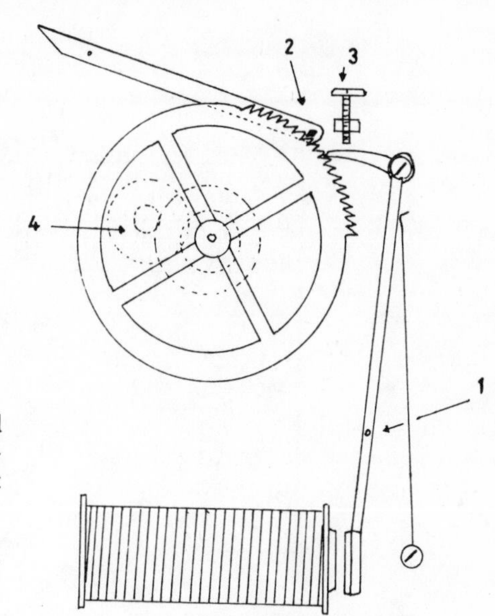


1. Sprung and pivoted pawl arm as armature of electro-magnet
2. Pivoted click
3. Pawl backstop
4. Minute and hour wheels

Fig. 32. Slave or dial movement of master-clock (*see* Plate 11)

the clock repairer. They are also of limited interest to the public now that cheap battery clocks will run accurately and for long periods without attention. Though fears used to be expressed that the mechanical clock would be ousted by the synchronous motor, we can now see that the mains synchronous clock has inherent restrictions—in particular the need for a nearby socket or the inconvenience of a long flex, quite apart from the chance of mains power cuts—compared with other clocks past and present. Its very silence can be a nuisance in that even a seconds hand gives less clear warning that a clock has stopped than does cessation of its ticking. Hence many synchronous clocks have been provided with a simulated 'tick' facility to be switched in if required.

Synchronous clocks may be started manually or be designed to start themselves when switched on, and their motors may be of high or low speed. Low-speed self-starting motors are general in modern clocks, but the other types are still often met and in fact the manual-start will show better when there is a power cut or other interruption, since it cannot start itself when current is restored.

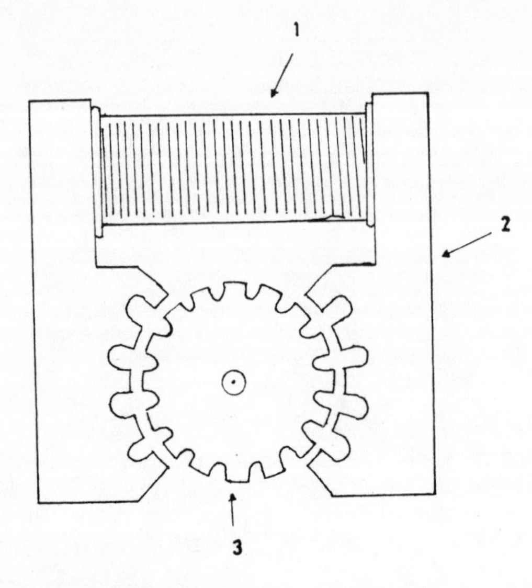


1. Field coil, connected to 50 Hz. mains supply
2. Laminated iron stator core
3. Magnetic rotor

Fig. 33. Basic multi-pole slow-speed synchronous motor

The essentials of all these motors are a magnetic armature and an electro-magnet, the stator or field coil (Fig. 33). The stator is shaped so that it encloses the rotor save for a small gap, the edges being the poles of the magnet, and it is made of stacked iron laminations wound with copper wire coils. The rotor has no coils, being a disc with projections corresponding to the poles of the stator. In operation the magnetic flux passes from one pole across to the other, forcing the rotor into the path of least reluctance. At a certain speed of the rotor this path will be established as well as when the rotor is not moving, and it is only at this speed that the synchronous motor will run. In the common arrangement, a rotor with thirty teeth runs at 200 r.p.m. with a mains frequency of 50 Hertz. The speed of a given rotor can be calculated from the formula:

$$\frac{\text{frequency} \times 120}{\text{number of poles}} = \text{revolutions per minute}$$

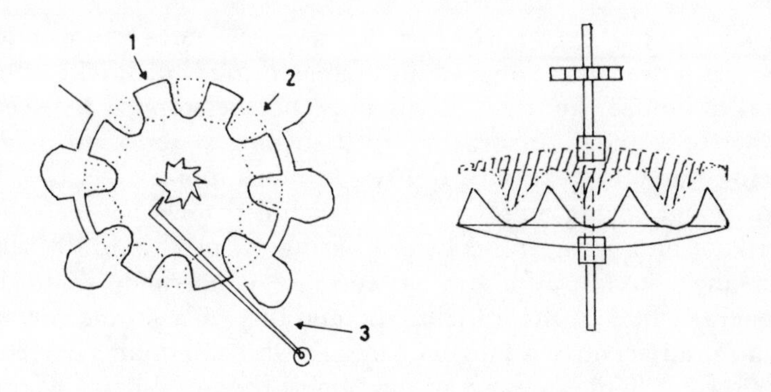


1. Magnetic rotor aligning with stator poles
2. Iron spider (shown dotted)
3. Pawl for ratchet wheel to ensure motor starts in correct direction

Fig. 34. Slow-speed self-starting synchronous rotor with spider

There are several self-starting arrangements. The commonest type employs a reversed rotor or 'spider' of soft iron immediately between the poles of the rotor itself, which is a permanent magnet (*see* Plate 12 and Fig. 34). Thus the rotor teeth are attracted to the poles of the stator when the motor is at rest. When the current is switched on, in effect the rotor teeth are repelled from the stator poles with the result that the 'spider' poles are attracted, and so the motor gets under way. The motor could then start in either direction, depending on the phase of the current when switched on. A loose pawl is therefore mounted against a ratchet on the rotor arbor so that the motor can start only in a direction which will turn the hands clockwise. Alternatively to the 'spider', the high-speed motor is made self-starting by 'shading' the poles with copper so that the magnetic field is delayed at each cycle in the areas which are shaded (Fig. 35). The result is a revolving magnetic flux which sets the rotor in motion.

There is no longer great demand for new striking clocks which are not of a traditional nature, but striking and chiming work has been used in synchronous clocks. The difficulty arises, of course, from the fact that a synchronous motor must not be slowed by taking on an increased load, or it will stop, and of course the main motor must not be stopped between times of striking. In practice rack striking systems are used, being put into gear with the synchronous motor by a clutch device employing an idler wheel which is thrown between the striking and going trains by the setting-off mechanism of the striking. So far as alarm systems are concerned, resort is generally had to the 'buzzing' potentiality of a simple spring placed adjacent to a magnet through which *ac* mains current is flowing. Sometimes an adaptation of the usual rising of the alarm wheel against a cam is used to complete a circuit containing such an alarm. More ingeniously, the magnetism already present in the stator magnet of the motor is used—the 'buzzer' would be operating all the time but for an obstruction mechanically placed in its path and removed at the appro-

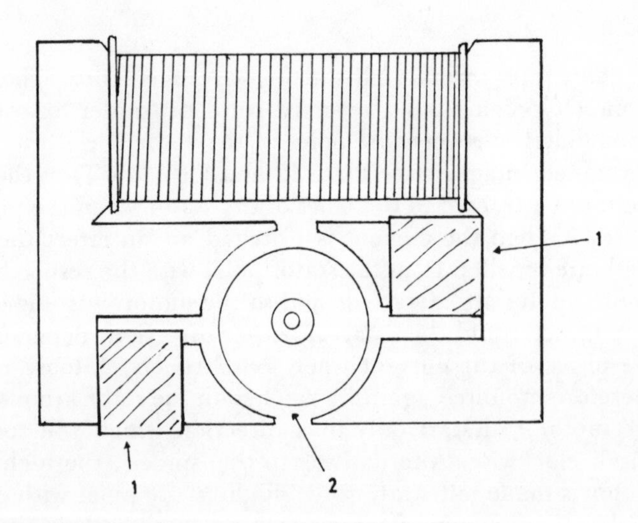


1. Poles shaded with copper rings
2. Steel disc rotor

Fig. 35. High-speed self-starting synchronous motor with shaded poles

priate time by a cam working with the alarm wheel (*see* Plate 12).

*Electrically Re-wound Clocks*

A third large group of clocks is that comprising those clocks fitted with the various forms of re-wind mechanism. These are basically conventional mechanical clocks, most often with light pinwheel escapements but sometimes with pendulums, whose mainsprings are rewound by electro-magnets or motors which the clocks themselves switch on. (On the whole, electric winding of weight-driven clocks is outside the sphere of domestic horology, though devices do exist for winding long-case clocks with electric motors.) The refinement of this type of clock, which could be produced cheaply and would run on one battery for a year, foreran the major development of transistorized clocks, and the type is now virtually obsolete. However, many are still going and some may still be found on sale, and these clocks may be nearer to the clock enthusiast's interests than the next generation of more 'electronic' clock.

Re-wind movements usually employ a mainspring which will drive the train and escapement for only about half a revolution of the winding wheel—this minimizes the power required from the battery and also ensures that there is little variation in power at the escapement whether the spring is fully wound or just about to be re-wound. In the basic conception, of which there are many variations, there is a small mainspring of conventional type mounted in a barrel (Fig. 36). This is not a going barrel, but a barrel whose only connection with the train is by a pawl mounted on one side of it. This pawl engages with a fine ratchet on the greatwheel, next to the centre wheel, and, provided that the spring is wound, pushes the wheel and drives the train. Alternatively, there is an arm or cam mounted on the arbor of the greatwheel and to this arm is attached a coiled wire spring fixed at the other end to a point on the movement plate—this spring is stretched when the clock is wound and returns to its natural length as the time for re-winding approaches. On the barrel, or on this alternative arm rotating with the wheel as the clock runs, is a contact, which gradually approaches another contact at the end of the sprung armature of an electro-magnet (similar to the contact

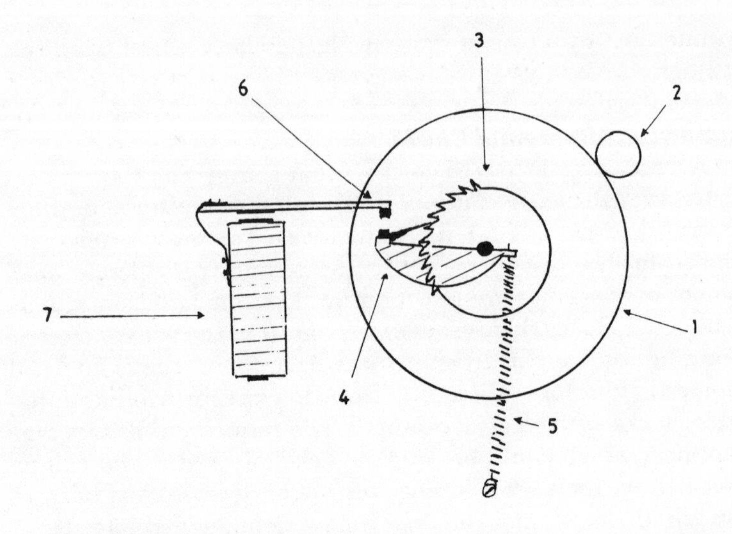


1. Great wheel, which drives centre pinion and on which is mounted the ratchet wheel
2. Centre pinion
3. Ratchet wheel fixed to great wheel and arbor
4. Spring pawl and contact on (shaded) metal section which is mounted loosely on the great wheel arbor and attached, at the far end, to the mainspring
5. Coiled wire mainspring which draws the contact piece towards the armature, thus also pushing the ratchet round by means of the pawl, so driving the train
6. Contact on extended armature of electromagnet
7. Electromagnet

Fig. 36. Spring re-wind by electromagnet

and armature assembly in the Synchronome clock).

When the contacts meet in the course of the spring's unwinding, the electromagnet's circuit is completed and its armature jerked to the magnet pole. This action forces the barrel or lever, against the spring (which is thereby re-wound) back as far as it will go. The barrel or lever does not of course push the train backwards since the whole action is extremely swift and the only connection with the train is the light click-work. The armature is so sprung that once the spring is rewound, the armature jumps back and contact is broken; if the springing is too weak or the contacts are sticky there is of course the danger that the circuit will not be broken and the armature contact will obstruct the lever or barrel as it again starts to revolve and the battery will be discharged.

By similar methods small electric motors are switched on to rewind, by means of a worm gear, longer mainsprings in other movements of this type. The details vary widely. Sometimes there is again a direct contact, sometimes there are contacts in a sealed tube which are activated as a magnet approaches them (known as a reed switch). In others the control of the motor is by means of an arm which at a certain stage releases one of the commutator brushes, allowing it to fall onto the commutator so that the motor will run. Where there is a motor, the mainspring is usually a long coil of steel wire round the greatwheel arbor. There is no clickwork, for the effect of the worm gear is that the motor can turn the winding wheel but the winding wheel (when wound) cannot turn the motor and the energy of the spring must therefore be directed into the going train instead. Such a mechanism may be extended to operate striking as well, an extension of the rackhook being used to keep the motor switched off until the rackhook is raised by the lifting piece for striking, when the motor, whose switch is sprung to the 'on' position, is energized and drives the striking train. It also winds the going train, by appropriate gearing and a clutch which prevents overwinding and stopping the striking.

A rather ingenious form of re-winding (*see* Plate 13) employs not a mainspring but magnetic power to drive the train and its light escapement (Fig. 37). Here again a small motor is used and it is switched on and off by manipulating one of the commutator brushes. However, the determining factor here is not the position of the winding wheel, but of the armature, which is mounted on a taut wire and can slide in and out of the field magnet, in this case a powerful cylindrical magnet. Because the core of the armature is iron, it is magnetically drawn into the cylinder, and this action, by means of a worm gear acting on a wheel, drives the train. When the armature is virtually inside the cylinder, and therefore the clock is nearest to being 'unwound', the armature

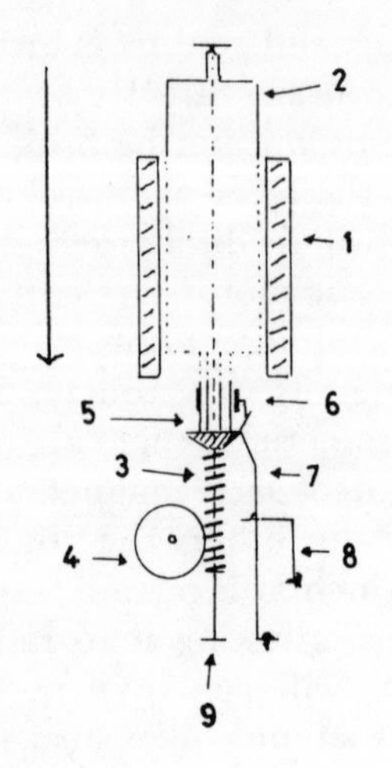


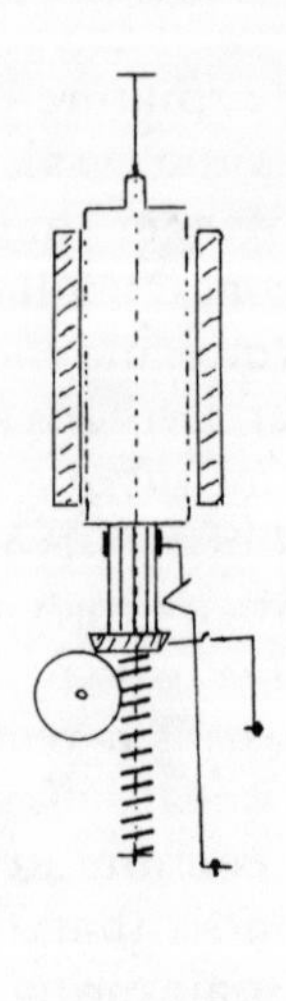

A. Fully wound

B. At the moment rewinding commences

1. Cylindrical permanent field magnet draws the armature into its centre, driving the train through the worm on the armature arbor
2. Armature with hollow arbor sliding on steel spindle (9)
3. Worm acting as rack when driving pinion (4), revolves only when rewinding
4. Pinion of going train driven by worm
5. Commutator
6. Commutator brush pushed away from commutator by cam and held away by hook (8)
7. Spring strip acted on by cam and pressing brush away from commutator
8. Hook holding brush away until nudged aside by cam when clock is unwound (B)
9. Steel spindle passing through worm, commutator and armature, which slide along it

Fig. 37. Motor rewind of magnetically powered movement (*see* Plate 11)

releases a commutator brush which, making contact, completes the circuit and starts the motor. As the gearing arrangement is such that the motor cannot drive the train, the motor drives the armature out of the cylinder and so quickly 're-winds' the clock. It is a fascinating mechanism to watch and is indeed reliable, but the brush switching apparatus is delicate and the movement is subject to dust and dirt in the long run.

*Magnetic Balance Clocks*

These are clocks with small battery-driven movements running on a single cell (1.5v), the balance wheels usually having on them one or more magnetic blocks which swing closely over or around small fixed coils without metal cores. They have recently gained a large share of the electric clock market, and indeed a considerable share of the total domestic clock market.

The principle is most easily understood from a version now superseded (*see* Plate 14, bottom). In this the balance staff carried an insulated plastic cam from which projected a contact pin which in the course of the balance's vibration struck a fixed springy contact. On the balance's rim were two coils, diametrically opposite, and above them, mounted on the frame, two semicircular permanent magnets. When the circuit was switched on by the contacts, the coils, and so the balance, were attracted and repelled by the poles of the permanent magnets, thus giving an impulse. The oscillating balance drove a small pinwheel with shaped radial pins (identical to those now used) by means of an interrupted worm shape, formed of metal slips, attached to its arbor.

This arrangement has the ingredients of modern magnetic balance movements, but the disadvantage of a mechanical contact, which disturbs the balance and is subject to dirt and sparking. The device which superseded direct contacts is the transistor. There is no space for a detailed explanation of transistors here; suffice it to say that in basic form they are minute electronic units capable either of power amplification (switching) or of oscillation—that is, of amplifying a minute variation of current, or of emitting regular low frequency alternating current when direct current is applied to them in a suitable electrical circuit. Both these qualities have been used in magnetic balance movements, but the function of switching (or relay) is the more often used, replacing the direct contact switch.

A typical movement (*see* Plate 15 and Fig. 38) has moving permanent magnets, fixed to the balance rim, and, between them, a very fine double coil of copper wire. There may be two coils wound together, or a single coil with a centre tap, according to the circuitry used. The function of one coil (or half of the single coil) is to receive the minute currents induced in it as the balance magnets pass over it and to feed them back to the transistor circuit. The function of the second coil (or half) is to return the amplified current in the form of a magnetic impulse to the balance. The transistor will operate as a switch only when current flows in one direction, so that the balance receives an impulse at alternate vibrations. The circuits are mass-produced modules and will be adequately energized for at least a year by an ordinary 1.5v dry cell, or longer with a specialized cell. The balance is regulated in the usual way by an index, often with a vernier scale, on the balance-spring, and it drives the small gear train by means of the metal (or nylon) slips of 'worm' section already mentioned. Alternatively, and with a higher current consumption, the transistorized circuit may produce a low frequency oscillation in the region of the natural period of vibration of the balance and balance-spring and feed it by means of a single static coil to the balance magnets—the balance is in this case set off-centre so that the oscillations immediately precede the arrival of the magnets over the coil, and regulation is more limited.

These movements are cheap and reliable. Once regulated, they keep very good time (though, unlike the rewound type, they lose time as batteries run down) and require little attention. Should they develop faults, they are replaceable, and often exchangable with models from different makers; repair is frequently not economic, the costs of obtaining a part or unit being close to those of fitting a whole new movement. They are also adaptable, being fitted into expensive reproduction cases or cheap plastic containers. Increasingly they are found as the 'works' of a clock—such as a Four Hundred Day 'electronic' clock where the traditional rotating pendulum is kept going by a flick from the train of a 'magnetic balance'

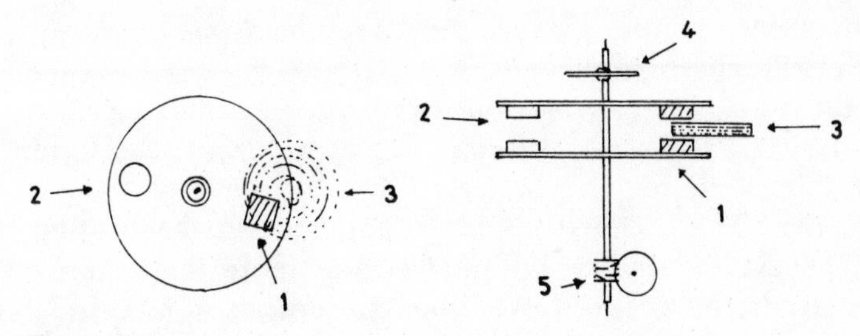


A. Variation

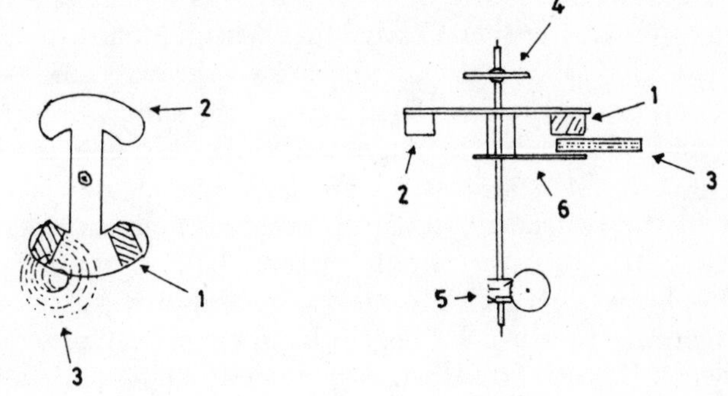


B. Variation

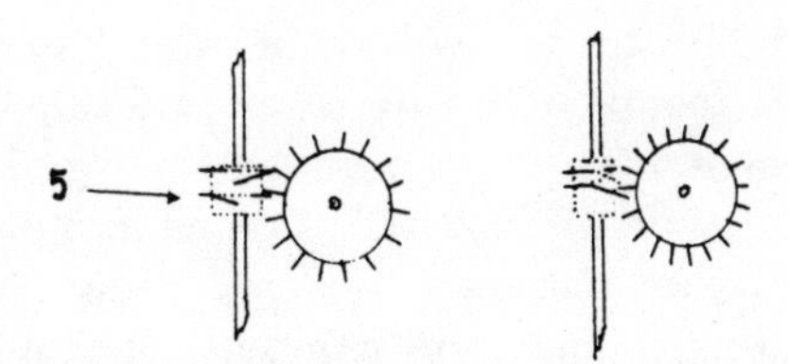


C. Details of spirals driving pin wheel

1. Magnetic blocks mounted on balance wheel, passing over/ round coil
2. Counterpoises
3. Coils receiving induced current from balance magnets and giving it, amplified by the transistor circuit, as impulse to balance
4. Balance spring
5. Interrupted worm driving pin wheel
6. Iron keeper for magnetic block

Fig. 38. Forms of movement with magnetic balance and energized coil (*see* Plate 15)

movement—whose appearance might lead one to expect a very different mechanism.

Provided that its vibrations can be divided into a beat perceptible by the senses without loss of accuracy, the faster a reliable oscillator the better, if it is to be used for timekeeping. The magnetic balance has been used vibrating at 36,000 cycles an hour, but in electronic terms that is not very fast, and a balance wheel is inherently prone to inaccuracies, arising particularly from the variable friction at its pivots. The oscillator of the quartz clock vibrates, as we shall see shortly, many times faster than any balance wheel. Meanwhile, friction is dispensed with in the oscillations of a tuning fork. A tuning fork can be kept in motion in a way very similar to that of the magnetic balance; it is magnetized, and one pole is extended inside a double coil of wire (Fig. 39). The magnet, moving in one of the concentric coils, induces a small current, and so causes the transistor circuit to send back amplified current through the other coil, this serving to 'impulse' the tuning fork, to keep it vibrating. Tuning forks in clocks usually vibrate at between 100 and 300 Hertz—giving out a note in the region of Middle C on a piano.

As is fairly well known from advertisement of the Bulova tuning fork watch, a tuning fork's vibrations can drive a train mechanically by means of a tiny pawl and ratchet. Another arrangement, akin to the synchronous motor, is also used. Here, Fig. 39, one end of the fork is shaped into two poles with a narrow gap between them. Between these poles there is therefore a magnetic field, and it moves from side to side as the fork vibrates. In this flux is an iron rotor with 'teeth' corresponding in shape to the ends of the poles and with inner teeth or bars, and the magnetic force travels via the path of minimum reluctance from teeth to bars and back. The rotor's speed is determined by the number of teeth which it has and by the frequency of the tuning fork, though regulation is available from a screwed iron block which can be raised or lowered in the field. The tuning fork starts to vibrate as soon as the transistor circuit is complete, but the motor is not self-starting; a lever is provided with which the first wheel after the rotor can be started and the rotor will then be jerked into motion, settling down quickly into the speed at which alone it can run.

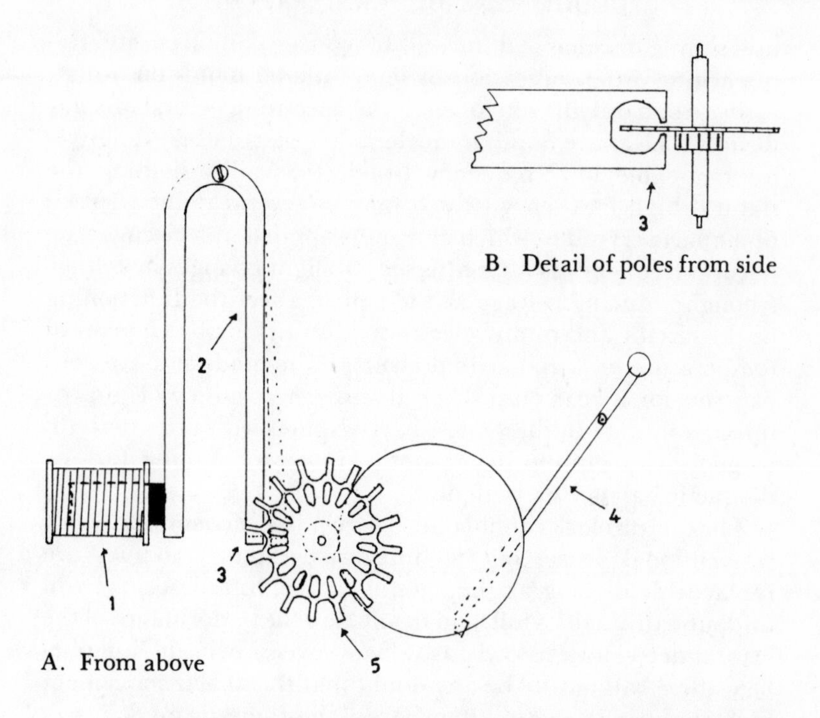


1. Concentric coils receive induced low current from tuning fork magnet and feed it back amplified as an impulse
2. Tuning fork magnet. (Vibration shown dotted.) Extends as core inside coil
3. Poles of tuning fork with which inner or outer 'teeth' of wheel tend to align
4. Pivoted starting lever to flick wheel
5. Soft iron rotor wheel

Fig. 39. Form of tuning fork movement

*Quartz Clocks*

The quartz crystal clock which at one time virtually filled a room and was the most accurate means of measuring time then in existence has now, as a result of developments in electronics, been reproduced at an acceptable price and in a sufficiently reduced size to replace conventional clock and watch movements. The quartz movement is essentially electronic and outside the sphere of the mechanical horologist, though it may use simple mechanical means to convert an accurate frequency to visual indication by conventional hands on a dial, rather than relying on one of the electronic visual display devices such as are familiar on electronic calculators. It derives a practicable low frequency by electronically dividing the natural high frequency (which may be several million Hertz) of a quartz crystal to which current is applied in a circuit. The natural frequency is little affected by slight changes in voltage (though reducing voltage will of course affect the functioning of the circuit and of any electronic display), or by changes of temperature or barometric pressures. The modern movement can run for a year on a dry cell with an accuracy of plus or minus a minute in the year. This compares with rates perhaps as good as a minute in a month for a good tuning fork or magnetic balance movement.

The quartz clock cannot often be 'repaired' economically in a conventional sense, but its integrated circuit modules are replaceable so long as they are available; 'obsolescence' will undoubtedly take its toll, and in any case the performance of the crystal deteriorates slowly as it 'ages' over a period. Nonetheless, there can hardly be any doubt that the quartz movement will take over an ever greater share of the domestic market, and in doing so become relatively cheaper still, so long as ancillary functions like striking and alarms (which consume more current) are not required to accompany the basic timepiece. There has never been so accurate a clock available so cheaply to the householder. Its availability seems to mark a decisive point in the development of domestic clocks. We may be seeing the first stage in the termination of the long evolution of clocks (mechanical clocks) as we have known them, so that in future they will be retained primarily for their interest as novelties or antiques, not for their accuracy or convenience.

*Chapter Six*

# CASES, DIALS AND HANDS

## *Clock Cases*

The varieties of clock cases through the centuries are of course virtually innumerable, although several fairly authoritative works on antique clocks go a long way to indicate what is typical of a particular period or area. Repairing cases is on the whole a matter of common sense or of skilled craftsmanship—part of the common sense lying in knowing what to hand over to the craftsman and what to attempt oneself. Nonetheless, it must not be forgotten that cases affect not only the appearance of clocks, their primary purpose is to exclude dust and, whilst one can go unnecessarily far in this direction for a movement of low quality, it is an aspect which must be attended to.

As an illustration of this point, many older cases, until at least well into the 19th century, had metal or wooden frets in their sides, and these were backed with silk. It has been maintained that all the movements with glass panels to show 'the works' (all, that is, of substantial old English clocks, not of later or foreign ones) were originally fitted with such frets rather than with glazed panels. The silk was intended at once to let out the sound of the striking and to act as a dust-filter. Certainly it attracted the dust. Comparison of the colours of the tattered remains and of the covered areas shows much fading, and frequently there are large holes which let dust in as well as letting sound out. Owners will insist with some justice that they do not wish for gaudy new silks, but it really does not make sense to clean and overhaul a valuable clock and then to leave an entry for all kinds of dust and dirt, most types of which are more damaging than in the days when the clock was made. From the same point of view, cracks and gaps in cases must be filled so far as this can be done without spoiling the appearance.

The filling-in of gaps, replacement of corner fillets, strengthening of backboards with battens (or replacing them), and the levelling of feet, or removal of an incomplete set which cannot be made good, all amount to structural repairs essential to a pendulum clock, for any movement of the case in sympathy with the pendulum will upset the beat and may even stop the clock. The same applies to the fixing or placing of movements in cases, particularly of large bracket clocks and long-case clocks where the shaking caused by the release of the bell hammer can be considerable and cause a movement to wander. The long-case movements of old thirty-hour clocks are sometimes placed freely on the cheekboard sides of the cases; this is not a very satisfactory arrangement, but the least that can be done is to make sure that the cheeks are sound and level, cutting them back and fitting any new wood if necessary. Eight-day and later movements are usually screwed to stout seatboards, either by screws directly into the movement pillars or by screwed hooks which fit round them, and the seatboard may be screwed to the cheeks. This can be inconvenient from the point of view of a need to remove the movement from time to time, and there is something to be said for using steel pins, without heads, rather than woodscrews for this purpose. Drive in the pins, or make them a close fit in the cheeks, so that holes in the seatboard drop over them. Bracket clock movements were mounted similarly until the 19th century, the seatboard being made with sides so as to form a little table. This is often fixed neither to case nor to movement. It is meant rather to support the forward weight, being supplemented by two strong right-angled brackets fixed to the back plate and through the sides of the case by threaded bolts, though many older bracket clocks are fastened to the seatboard through the pillars in the same way as long-case movements. Later, when movements were less heavy, the mounting was transferred to lugs screwed or riveted to the four corners of the front plate and screwed to the inside of the case front. This change corresponded in part to the more widespread use of round dials and bezels, whereas old solid square dials were designed to fit closely into the front aperture of the case behind the opening wooden door, and took some of the weight. These lugs are a common source of trouble, since the wood of cases may not be thick or hard enough to fix them securely and they are sometimes not mounted rigidly on the front plate, so that movement is possible. It is sometimes necessary to increase the number of fixing points for the lugs, on case or plate, or both.

The finishing and restoring of old wooden cases is a matter of taste and of the skills available. Marquetry and inlaid work, and also lacquered 'cases', are best left to the expert if a job of any size is involved, and cases of these sorts are usually quite valuable, if not to everyone's liking. Raised veneer on a polished case can be separated with an oblique cut from a razor blade and glued back into place; where the raised section is in the form of a bubble it should be pierced with a very fine needle, then pressed into place with a warm iron, with additional glue inserted if necessary. A polished case with a good patina can be treated with a stiff mixture of beeswax and real turpentine, having been cleaned if necessary with a solution of turpentine, linseed oil and vinegar. There are occasions when a good old case has been painted over with a heavy varnish, or stain mixed with polish, so that the result is soupy, but other than in such instances good taste is surely against totally stripping and then oiling plain wood which from the first was surfaced and coloured. Mirror-polished shellac finishes can be cleaned by wiping with a wad charged with oil and methylated spirits and then touched up with button polish, but the total repolishing of a french-polished case of any size is best not tackled unless one has gained experience in this area. Most wooden cases have reliefs in the form of brass keyholes, hinges, bezels, finials and the like, and the general effect is enhanced if these are thoroughly cleaned and given a coat of clear lacquer for protection. Replacements are available, though often not in the identical style. Where there is some indication that wood or other ornament was originally gilt, it can usually be touched up with a gold wax or liquid leaf, but pieces of any size are better sent away to be gilded.

With regard to cases and movement-fixture, French clocks are generally in a category of their own. The French adopted the round hinged bezel relatively early and the standard arrangement is to have two such bezels, one at the front and one at the back. The rear bezel may be glazed or be fretted in metal and with the usual silk backing. The two bezels are linked by lugs riveted or screwed to the front, into the ends of which long screws are fitted from the rear bezel. Thus in effect the movement and bezels clamp the case between them. The movement is screwed or pinned to the front bezel by means of short pillars in the plate which reinforces the dial. The securing is effective generally, but there is the disadvantage that the strain of winding imposes a torque on the movement which tends eventually to cause it to become loose, so that the clock goes out of beat and may well stop. If, on the other hand, the fixing screws are excessively tightened, they are liable to strip the threads of their holes in the fixing lugs, or the lugs come adrift from the front bezel. It is not, however, difficult to make new straps or if necessary to retap the holes in the old ones to take fixing screws of slightly larger diameter.

The cases into which these rather standardized (but usually highly-wrought) movements are inserted come in a great range of materials (gilded metal, wood, slate, alabaster, marble, plaster) and shapes (elephants, plain boxes, triumphal arches, reclining figures, mythological beings and beasties, and so on). There is no accounting for tastes and one man's meat in this matter is another man's poison. If one is presented with a clock to repair which appears simply grotesque, one simply has to do one's best by the material of which it is made and to pay tribute to the variety of human nature.

Where wood is concerned, the treatment is as described above. A common late finish is ebonizing. True ebonizing on hardwood shows white when chipped or scratched, since plaster is a large component. It can be filled with a filler and touched up with a mixture of pure french polish and black aniline dye. Many apparently ebonized cases are, however, produced by staining black and following with layers of french polish. At all events, the temptation to resort to a spot of black gloss paint should be resisted, because it will harmonize with neither finish. Marble, slate and alabaster cases respond to a thin cream of turpentine and beeswax, when they have been cleaned, if required, with metal polish (which will remove the surface) or a proprietary car 'colour restorer' (which, again, is finely abrasive). They can be filled or repaired with plaster of paris or a modern resin filler, with metal supports if required. These often massive cases are frequently relieved with engraved patterns which were once gilt, and they can be restored with liquid leaf or, for a less bright (and less durable) effect, gold paint. The same remarks apply to bezels, finials and ornaments as above, except that ornament is more often gilded in French clocks. Where the whole case is of gilded metal, the amateur has little chance of rejuvenating it, though it will improve with a wash of warm soapy water. Metal polish will make matters worse as it will remove the gilding. These cases should generally either be washed and left as they are, or sent away for regilding—which may result in too great a brightness for the owner's taste. Restoration wax does no harm.

Carriage clocks are a special class of predominantly French clock—the highly sought-after English carriage clocks are similar as regards their cases, but there is more variety of shape and material. These clocks are basically of brass, which may be polished and lacquered, or gilded, and with glass panels on all sides, though the doors may be metal and the panels may be of valuable coloured enamels. Wooden cases are also found on clocks which would be classed as carriage clocks, particularly with English examples. Entry is through the base, where a brass cover (often missing) obscures four screws into the pillars. At the other end, these pillars may be screwed into the case-top in one of several ways, or the top may be solid with them (as is usual in older clocks, where also the handle is usually sprung between cast lugs rather than screwed on). The door is most often a glass panel slotted into a brass frame but, again mainly on older clocks, may be of chased metal. The movement is usually secured to the base by two or four screws into the movement pillars, but alternatively small screws are driven into the thickness of the plates themselves. Sometimes there is a plate screwed to the case pillars and holding their glasses in place, but more often the glasses will fall out when the movement and base are removed.

Many of the more elaborate cases with fine chased work were gilded, and at the present cost of gilding and the constant demand for these pieces, it is usually worth having the same finish applied again. There is no need, however, to gild the plainer items and it is usually clear that they were not originally gilded but merely lacquered. They can be cleaned with the rest of the movement, rubbed up carefully with metal polish and then warmed very slightly to help in the application by cottonwool of a coat of clear lacquer. Owners should be warned that the cases must not then be cleaned with metal polish or the lacquer will be rubbed off and the finish will very soon deteriorate.

Replacement glass panels for these clocks are generally available from materials suppliers, though there can be difficulty in obtaining the curved glasses for oval clocks—the only solution is to wait and to keep searching. It not infrequently happens that the panels are loose either because of incorrect size or because of distortion in the metal of the case. Even if such glasses stay in, they cause nasty noises in striking clocks and permit an entry for dust. Loose sides can be wedged with slips of cork or matchstick pushed into the slot—sticking with adhesive is rarely successful, often unsightly, and will cause problems when the clock is next dismembered. Distorted cases must be corrected, for they cause cracked glasses when screwed up. The doors vary in fit and are often too tight for convenience or too loose to serve their purpose. Sometimes they can be corrected by bending the case pillars or screwing them in a slightly different position, but it is more often a matter of replacing the hinge pins, which are driven into the back pillars of the doors, and, possibly, of moving the holes into which they fit. In bad instances, some metal can be removed if the appearance is not thereby spoilt.

*Dials*

The restoration of dials has to be approached with care by the amateur, mainly because of the instability of the markings, which will often rub off if drastic cleaning is attempted and on some types are very difficult to restore. The figures of old brass dials with separate chapter rings and spandrels are, however, fairly easy to fill in with a black paint or wax, since they are wax-filled engravings and the line is clear. Brass-work of such dials can be cleaned to taste and the chapter rings can be resilvered in the home by the traditional process, which is described in a number of manuals. They can, however, sometimes be brought up considerably, without going to the bother of resilvering, by rubbing with a paste of cream of tartar and water. Cleaned and resilvered rings must be thinly lacquered or they will soon deteriorate. Complete sets of spandrels can be bought but will rarely match the originals; the alternative where a spandrel is missing is to make a mould from the others and to cast a spandrel in resin, or lead, and then to treat it with liquid leaf or restoration wax until it matches the colour of the others.

Deciding on the best approach to chipped or illegible painted and enamel dials is very much more difficult. The original tint, modified by time, is as a rule very difficult to match with white cements, although of course the attempt may be made to rectify damage in this way and indeed, if a match can be obtained it is sometimes even possible to scrape back a crack so that it no longer appears as a dark line, and to fill it almost invisibly. In general, however, cracks are best left alone. Markings on painted iron dials can be professionally restored and a very attractive appearance may result, though the danger is always of over-restoration. Small blemishes can be made good by painting with oil paint, but correction of markings and figures, which are often very fine, as also the filling in of missing patterns, is difficult since there is no engraved recess to act as a guide. The restoration of carriage clock and French clock dials, which are glazed enamel on thin copper and extremely fragile, is generally impossible, though cracks and chips can sometimes be touched in without harm with porcelain paint. The severely damaged dial which is simply intolerable can be covered with a photographic replica, available from suppliers, or the replica may be mounted on a substitute dial plate and used without the original. In such cases, care has to be taken that the hands completely clear the dial hole.

*Hands*

Hands can virtually always be replaced, but not always with pieces of identical pattern; identical hands can of course be made or adapted from a suitable pair, but it is a very long job and not often worth while. Broken hands may, according to their fineness, be repaired with brazing or silver-soldering, sometimes with a support placed behind the join. (Soft solder will not stand the strain where the hands are set from the front.) Replacement hands, from whatever source, may well not fit the arbors as required, but it is usually possible to remove the collets of the original hands, or even the bosses if there are no collets fitted, riveting or soldering them, according to type, on to replacements. Collets with square and round holes can be bought in assorted sizes. On clocks of any age, the use of hands of appropriate and preferably original design is important; the styles of hands used at particular periods are indicated in works on antique clocks. Replacement hands can often be detected by the fact that they are not of exactly the correct size; in a true set of hands, the hour hand tip just reaches to the feet of the hour numerals and the minute hand tip passes across to, but not over, the outermost ring of the chapter, thus aligning with the minute divisions if these are marked. Old steel hands are usually blued. Blueing is a process of heating the perfectly bright cleaned steel on a brass plate until it reaches the desired shade of blue, and then plunging the metal into oil. It is practicable in the home, although several attempts may have to be made to produce the even colour resulting from a proper distribution of heat. Heating in blueing salts, quenching in water, is also very satisfactory.

The fitting of hands varies considerably as to period, and indeed within periods. On older clocks where the hand-setting (Fig. 7) is by means of a frictionally-fitting cannon pinion or a cannon pinion with a spring between it and the front plate or the arbor, the minute hand fits on to the square end of the cannon pinion and is secured by the pressure of a washer (collet) and pin passing through a hole in the tip of the centre arbor. This pressure has to be carefully adjusted so that it is even wherever the hand may be, and so that it is neither so excessive as to reduce the power available to the escapement nor so slack that the hand's movement is unreliable or the pin falls out—the 'feel' of a properly tensioned hand is soon acquired by experience. The pin should fit well, the hand's tightness being controlled rather by the thickness of the collet. In a striking clock the hand must of course be so placed on the square that it is pointing exactly to 12 when the hour is struck; if, when properly fitted, it is not quite on 12, then it may be slightly bent or (preferably) the squared collet can be held while the hand is turned on it in the right direction, the collet then being riveted tightly to the hand with hammer and punch. More drastically, it may be necessary to remove the hour lifting pin. The hour hand collet may be a split bush or pipe fitting frictionally on to the hour wheel pipe in any desired position, or it may have a square hole fitting a squared pipe. In the latter case, there is often a small screw to fit the hand to the pipe. Again, the position has to be set to match the striking; if the fitting is square, this setting is done by turning the hour wheel slightly in relation to the minute pinion (not forgetting that if the snail is mounted on a starwheel, it can be turned to the nearest hour). Remember also that, in calendar devices operated by a pin on the hour wheel, matters must be adjusted so that this pin connects with the ratchet at around twelve o'clock (midnight).

In more modern clocks, with internal clutches and often with hand-setting from the rear, hands are not often pinned on. If there is hand-setting at the rear, there is no strain on the minute hand, and it may merely be pressed on to the arbor. Sometimes the arbor is squared and the hand is held on to the square by a nut screwed into the threaded tip of the arbor. This is common with modern striking and chiming clocks set from the front. The hour hand is seldom fitted with a square and is pressed on to the hour wheel's pipe at the appropriate place; setting to the required hour with a sounding clock usually means fitting the hand to correspond to the position of the snail and striking, for the snail is not usually mounted on a separate starwheel. If the minute hand arbor is squared, the same adjustments have to be made as mentioned above, but if it is a press fit to the arbor then it has merely to be correctly placed.

Alarm hands are almost always just pressed tightly on to their arbors, the sole requirement being correspondence with the initial setting of the clock. Thus the procedure is to set the clock until the alarm is just going off (or to move the alarm similarly) and then to press on the alarm hand so as to indicate whatever time the clock may be showing. Alternatively, press on the alarm hand as the alarm starts, and fit the other hands accordingly so that all hands show the same time. It is then wise to run one or two checks, for it is maddening to have set up a clock and returned it to its case, only to find that the alarm sounds at seven o'clock when its indicator specifies six-thirty. Centre-seconds hands, again, are simply pressed on to their arbors, which tend to be fine and fragile. It is rare with domestic clocks to envisage any use for centre and other seconds hands apart from being decorative and to show whether or not a clock has stopped. Hence there is very little point in elaborately ensuring that the seconds hand points to 60 when the minute hand records a complete minute, unless you propose to adjust it with equal care on every occasion when you set the hands to time, for the set hands device or clutch does not, of course, include the seconds hand and this will not be moved when the other hands are moved. It may be noted also that the pendulums of Vienna Regulator wall clocks do not beat full seconds, and their seconds hands do not therefore point step by step to the exact divisions of a normal sixty-division chapter ring. Once pressed on, the seconds hand of any clock should not be turned on its arbor, for there is a risk of damage to the hand and the escapement.

## *Chapter Seven*

# DISMANTLING AND CLEANING

Periodical cleaning of a movement is necessary partly for the sake of appearance, if the movement is visible through glass panels or under a glass dome, but principally to assist in long-term running. However well the case fits, in due course particles of dust and dirt will enter the movement from outside, just as minute particles are generated inside from the friction of engaging parts. These are abrasive in action and will work to destroy further the mating surfaces, especially when they are caught up in the oil, which when clean, assists in free running. For this reason, oil is applied only in minute quantities and as far as possible kept clear of all wheel and pinion teeth save for those of the scapewheel. Nonetheless, it is well known that clocks require oil, and many a movement reaches the repairer flooded in machine oil which is quite unsuitable and has generally been applied to every accessible surface. This oil and suspended dirt has to be removed, as have any particles of rust arising in unoiled and slightly damp parts. Tarnish is of little import on brass surfaces which do not actually engage, but it is removed anyway in the cleaning process and must be polished out of visible movements.

Since the effect of dirt is cumulative, it is obviously desirable to clean a clock whenever it receives attention if some time has elapsed since it was last cleaned, but it is a long job to do properly and this is not always practicable. The fact remains, however, that often critical faults are invisible until a clock is stripped down and there is nothing like taking a clock to pieces completely, preferably cleaning it, and reassembling it, for giving an insight into how it works and its possible troubles. Except for the question of appearance, there is no such thing as partial cleaning; the most important parts to be cleaned are the pivots and pivot holes, and these can only be cleaned with the movement completely dismantled.

### *Dismantling*

Before stripping a clock down, check it over for apparent

maladjustments and breakages—more on this is said in the next Chapter. The dial and motion wheels can then be removed without difficulty, but further than this one cannot go without letting down the power supply, if it is a mechanical clock. It cannot be assumed that merely because a clock will not 'tick' or strike, it is safe to dismantle. The trains and other mechanisms may well be obstructed, masking the fact that power is left in the spring. The power is of course removed from weight-driven clocks merely by taking off the weights, but for a spring-driven movement, with or without fusees, a more cautious approach is necessary. It is possible to run the movement down by holding back the locking piece and removing the escapement pallets, or platform escapement, but this is not advisable as a general rule because of the chance that there may be damage to a wheel or pivot already or that they may be damaged in the process. It also takes rather a long time.

With an alarm clock in which one spring drives both alarm and going train, there may be no choice but to remove the balance and lever and to let the movement run down. With most spring-driven clocks, however, it is best to let down the springs themselves. This is done by grasping the winding square, or the barrel arbor square of a fusee movement, with a large key or preferably a hand-vice, and moving the click away from the clickwheel, letting the key or vice rotate progressively in one's hand and taking the utmost care that things do not get out of control—the powerful springs of an old English clock with fusees are particularly dangerous, and the movement of any clock is liable to be damaged (as is the person holding it) if the spring is allowed to go free. When the spring is apparently down, remember that in addition stop-work may have to be removed as the power is held off, or the wheels will burst out and risk damage when the plates are separated—the same applies with a fusee in that the line or chain must be slack before the spring can be considered safely exhausted. It is as well to remove the platform escapements of carriage clocks before letting down the mainsprings, since these escapements are easily damaged; wedge the train or hold it by carefully tightening the contrate wheel screw, while letting down the springs. The same can be done with the relatively delicate crutch and pallets of pendulum clocks, or these can be removed directly the power has been let down, for they will only be damaged and may hinder later work if left on.

Most clock plates are held together by pins passing through the pillars. The very oldest are sometimes 'latched', with metal hooks slipping into slots in the pillars, and many modern movements are secured with screws into the ends of the pillars. There is no universal rule as to whether pillars are fixed at the rear and pinned at the front, or vice versa—the modern screwed movements indeed often have the pillars all screwed at both ends rather than riveted in. Thirty-hour one-handed movements follow old lantern clock practice as a rule and are 'posted'—there being four vertical iron pillars at the corners and a brass top and bottom plate which are fixed to the pillars. The trains are arranged with the striking train at the back, and the arbors run in brass strip plates, usually locating in holes in the bottom plate and being wedged with iron wedges into slots in the top plate. Continental clocks and clocks, including cuckoo clocks, of the 'Black Forest' type often employ a frame of hardwood into which are wedged wooden plates which may carry brass bushes for the arbor holes.

In stripping clocks it is best to follow an orderly procedure, for this speeds things up even if you know the type of clock well from experience. There is much to be said for laying the parts out on a sheet of white paper in the order in which they are removed, thus ensuring that they can be replaced in the reverse order, if necessary drawing diagrams on the paper and making notes of unusual features. Often the clicks and other common parts of the striking and going trains are specially marked, but if they are not it may be worth giving them a distinctive punch-mark, for even if they were once interchangeable the process of wear may have made them no longer so. It is essential to keep screws, studs and springs with the parts to which they belong. Where there are a great many arbors of wheels, they and other parts can be placed upright in an inverted cardboard box with holes in, to indicate which way up they were in the movement and their approximate position. It naturally makes sense to remove the external elements—clickwork, hammers if any, motion work, rack, countwheel, snail and so on—before separating the plates and taking out the wheels. Where there is known not to be a broken mainspring and there is reason to suspect trouble in the train as causing a loss of power, it is worth at this stage testing the freedom of the relevant wheels between the plates, when most of the parts are removed.

*Cleaning*

Cleaning is simple but time-consuming. It consists of giving all brass parts, or parts with brass in them, a long soaking in a solution of soap (or washing-up liquid) and household ammonia, a solution which is at its best when warm. The pieces must be completely covered in the liquid or unsightly high-water marks will result which can be very difficult to remove. They will not come out bright—in fact they may well look worse than before they went in—but the dirt will have been loosened and the oil removed and in fact a surprising amount of sludge will be found at the bottom of the container—for which an old baking tin or oil drip-tin, as sold for use beneath the car, is very good since it enables the parts to be seen. The brass parts have then to be dried and brightened, either by scrubbing with a fine brass brush (such as is used for cleaning suede shoes) or by cleaning with metal polish, or both.

The treatment of brass and steel parts depends on how well finished they are. If their surfaces are not polished there is no particular point in trying to polish them. If they are polished the aim must be to make them as bright and smooth as possible. All kinds of fine emery, buffs, steel wool and fine scratch brushes (never on an electric drill!), and so on, may be pressed into service. Rust must be scraped or filed away even if the resulting surface is defective, because it will only spread if left. On fine movements screws and small springs are often blued. As has already been mentioned, they must be made completely clean, bright and free from grease (including that from fingers). They are then heated to the desired shade on a brass tray before plunging into oil while hot. Be careful, of course, not to overheat springs, or they will be softened. Blue enamel is apt to appear heavy and unsightly on these parts and it is better to reduce them to bright steel rather than to give them a coat of enamel. It may sometimes be used effectively on hands, however, though care must be taken to keep the metal level until the enamel has dried, or there will be irregular colouring.

The plates and pillars should also be soaked in the cleaning solution, and they may as well be left in there whilst attention is given to the smaller parts already cleaned. When they come out, they should be rubbed with tripoli and oil, or with metal polish, according to whether a polished finish is required. Although it will be slightly spoiled later, this should be done before the holes are cleaned, otherwise abrasives from the cleaner and dirt from the surfaces will find its way back into the cleaned holes. Pivot and other small holes can really only be cleaned one by one with pointed pegwood, the wood being twirled in the hole, removed and scraped clean and then returned to the hole repeatedly until it no longer comes out with dirt on it. Larger holes are most easily cleaned with strips of wool-free material or chamois leather, one end being tied to a fixture and the other end held in the hand, so that the hole can be moved to and fro along the strip. The same method is effective in cleaning the crossings (i.e. between the spokes) of wheels. Wheel teeth can be cleaned with a brass scratch brush, and here a revolving brush—even if merely fitted up in a hand drill—is immensely useful. The leaves of pinions attached to wheels are not easy to clean, but a reasonable job can be done with flattened and sharpened pegwood, preceded if necessary by a corner of fine emery or by scraping with a fine screwdriver. The leaves and their corners must be completely dry.

There is no special problem in the cleaning of pendulum clock escapements, but balance wheel escapements require careful treatment. Unscrew the balance cock and gently prise it up at the back—a small slot is often provided at the point for the purpose. Then, keeping a hand over the cock and balance, invert the escapement so that the balance can fall out; make very sure, however, that it is free to fall, and move the scapewheel and lever (if any) as necessary to release it. Set the wheel and cock aside and unscrew and remove the lever, if it is a lever escapement, and scapewheel. There is no need to remove the 'chariot' in which the balance of a cylinder escapement is mounted on the base platform. When all the parts have been removed, the platform may be cleaned—it is not usually dirty enough to need soaking but can be dipped in benzine—and also the lever and scapewheel cocks and their screws. Care is needed here with the jewelled holes—these are best cleaned gently with pointed pegwood, but remember that they will crack if pressed too hard. The scapewheel of an old offset lever escapement is usually brass and will respond to metal polish. Cylinder and modern scapewheels (other than of pin-pallet escapements) are normally of steel. They are rarely rusty or seriously dirty and will come up with a little oil and buffing with a chamois leather. They are all, and particularly the cylinder scapewheel with its raised teeth, very delicate.

The balance spring is next unpinned from the balance cock and gently pulled out of its stud and the regulating index so that the balance is free of the cock. Ideally, the spring should be taken off the balance, but this may not be necessary if the balance is not very dirty, and is not recommended to the absolute beginner. Clean the balance itself for the sake of appearance, but the roller and very fine pivots with an eye to the fact that these polished surfaces are vital to the functioning of the escapement. A rusty balance spring cannot be well cleaned, and blueing a spring is not practicable, so replacement is the answer in a bad case here. Oil on a balance spring is, however, as serious as rust, for this will cause its coils to stick so that the action is unpredictable. A balance spring with sticky coils should be soaked in benzine or paraffin and then dried between blotting paper. The spring is removed from the balance by means of a thin wedge, such as the flattened blade of a fine screwdriver reserved for this purpose. This is inserted from above into the gap in the collet on the staff. If the tool fits correctly, it will open the collet slightly and the collet will grip it, so that the hairspring and collet can be taken off with a slight twist and pull. To attempt to remove the spring by the leverage of a knife or screwdriver beneath the collet is to risk breaking the balance staff and ruining the spring if the tool slips upwards, but it may still be necessary carefully to loosen the collet in this way first. Many bent balance springs can be straightened with tweezers and plenty of time. The commonest straightening job is the result of a balance which sticks during the removal process, as a result of which the spring acquires a helical twist which will press it against the wheel rim or cock when it is remounted; straightening is a matter of boldly pulling it from the centre in the reverse direction when removed, meanwhile holding the outside end. Practice on old balance springs will help to give a 'feel' for the task. Should all else fail, assortments of hairsprings, with or without collets, are available for replacements.

The structure of the separate balance cock varies slightly according to date and type (Fig. 40). If a good appearance is required it is necessary to remove the regulating index so that the top of the cock can be properly cleaned. In the older types of cock, the index was secured by being placed under a polished steel disc with a flanged edge, the disc being held down by two small screws beneath the cock. The disc had a hole to retain the edges of a jewel endstone on top of the balance staff pivot hole. This is the normal arrangement in cylinder escapement cocks. It frequently happens that the index, usually polished steel, is bent and loose, and adjustments here are well worth while when the pieces are taken apart for cleaning, which is a matter of the sparing use of metal polish. The indexes of these cocks are in one piece, but with a slit in the circle held by the flange, so that they are gripped friction-tight by the flanged disc, and if they are bent regulation will be haphazard.

In the cocks of more modern platforms the endstone and pivot hole are permanently mounted in a similar flanged disc which may, however, be heavily pressed into the cock and should not be moved—a damaged jewel hole necessitates, of course, outside attention. The index is again mounted by way of slit circles under this flange, but is in two parts, one being the index lever, and the other the index buckle or pins through which the balance spring passes. Each part has its circle and one presses tightly into the other, with the result that, though the pins normally move when the index lever is moved, the latter may be moved independently if the pins are held. This allows a greater movement of the index, which in the older one-piece type would foul the hairspring stud earlier. The arrangement must be snapped apart from below for cleaning or to straighten the index—it is never advisable to straighten indexes when the escapement is assembled as the risk of damage is too great. Steel indexes may be glass hard and break

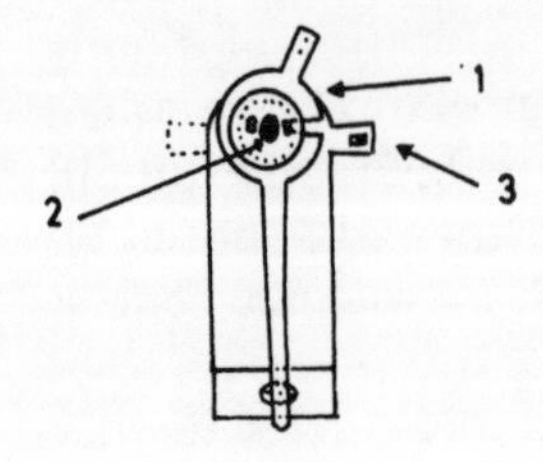


A. Older style

1. One-piece index and regulator clamped by flanged disc
2. Flanged disc holding endstone and index
3. Balance spring pinning block

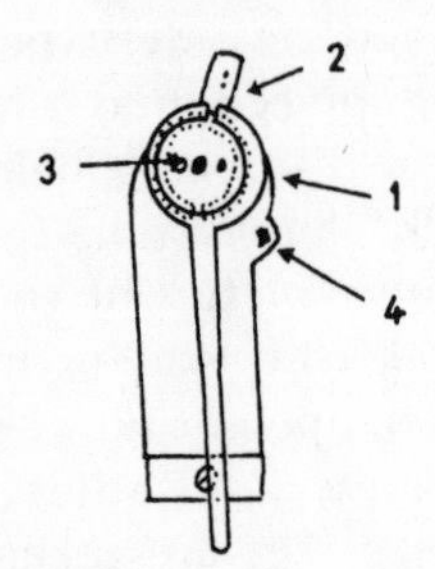


B. Modern style

1. (Outer ring) Index
2. (Broken ring) Regulator
3. Flanged disc (which may be friction jewel setting) holding index and regulator
4. Balance spring pinning block

Fig. 40. Indexes and regulators

if an attempt is made to bend them.

Pin-pallet lever escapements are often built into the movement rather than being of the platform type. They vary as to how much can be removed without taking the plates apart, but it is invariably possible to have the balance and spring out of harm's way at an early stage, and often the lever can be removed as well. Often there are not separate cocks for the parts, the balance being suspended on its conical pivots between two coarse screws with corresponding hollows for the pivots; the balance is removed by loosening one or both of these screws after the spring has been unpinned from its stud. The scapewheels are normally of brass and there are no special points on the cleaning of the escapement, save that the pallet pins (in some cases extremely fine) and lever shaft must not be bent in the process.

There is, for the amateur, no quick or easy way to clean a clock. On the other hand, he has the knowledge that there is in any case no better way than hand cleaning and polishing piece by piece. There is a sense of craftsmanship about a well-cleaned movement. There is also a feeling that the job was not done properly and that almost anything may happen in the case of a movement where the obvious repair was attended to, but the clock was not stripped right down; other faults may be latent, and friction and dirt will be slowly making their inroads into the mechanism. Yet stripping and reassembling a clock movement is not as complicated or as fraught with risk as it is sometimes popularly reputed to be. If you have acquired a clock needing attention, do not be daunted at the idea of taking it to pieces and thoroughly cleaning it. As far as possible, insist on cleaning movements rather than merely 'getting them to go'. Your satisfaction will be greater and the work far more reliable as a result.

*Chapter Eight*

# REPAIRS AND ADJUSTMENTS

We are concerned here with some of the commoner troubles to which movements are prone, and with how to rectify them. On receiving a clock for repair, it is advisable to make an initial diagnosis so that the likely area of trouble is identified before other parts of the mechanism are mistakenly altered, perhaps making matters worse.

## *Diagnosing the Trouble*

Assume first that a clock has come to you which simply 'won't go'. This is the situation where the pendulum or the balance swings freely, but there is no sound from the escapement and the hands do not move. First examine the hands. If they are crossed or rub each other, that may in itself be the trouble, as may a seconds or minute hand which rubs the dial. If the hands point to between twelve and one o'clock and it is a striking clock, on the other hand, then it is more than likely that the clock has failed to strike twelve correctly, for this, as has already been said, will stop many clocks (Fig. 20). Push the minute hand back to just before twelve and wind the striking—if it is fully wound, the clock may strike, and you will have to find out why it failed to strike in the normal course of running. If the strike appears to have been nearly unwound, it may well be that it has not been properly wound and that this is the cause of the failure. Either way, if the clock now 'goes', the source of the trouble is on the striking side which has interfered with the functioning of the going side.

Examine the movement, from the rear if possible, to see whether the scapewheel drops on to the pallets under power or is more or less static. If this wheel shows life, there may be a considerable fault in the escapement. If it does not, with your finger apply power to an earlier wheel in the train. If the escapement now moves freely when previously it did not, there is a failure of power either within the train, or the mainspring may be broken. You will know, from applying power to the earlier wheel, whether it was under traction from the spring or weight, whether it moved freely in either direction, or whether it was stiff. The latter condition points to a stiff pivot (or other arbor interacting with a wheel) either here or nearby in the train. Test the barrel itself of a spring-driven clock (or the fusee—having of course noted that the line from the barrel to the fusee is intact). If the clock is wound, the barrel will be difficult to move, but will spring back in one direction. If it is moving freely in both directions, the spring is almost certainly broken or disconnected within the barrel. If the barrel is stiff in both directions, not from the power of the spring, its arbor will be seizing, its teeth crooked and jamming because of a loose barrel cover, or (a not uncommon fault) the edge of the spring may be jamming on the cover, because the spring is too high, is bent, or the cover is badly tilted. Whilst your attention is here, examine carefully the state of the clickwork. If the click is unscrewed, broken off, its spring bent away, or if the teeth of the ratchet clickwheel are worn or broken (as is common with a click too thin for its wheel), the whole power of the spring can be released and the movement will then be in the same condition as if the spring had broken, although it may still wind up partly. Should such a release have occurred, bear it in mind as the cause of the trouble when stripping the movement and cleaning. In such a case it is never sufficient to leave the movement alone and merely to attend to the clickwork, for the suddenly released spring nearly always breaks wheel teeth or pinion leaves, burrs the barrel teeth and bends arbors, leading to stiff and intermittent running. The spring may also break itself or come disconnected from its hook in the barrel or on the arbor. Such faults seldom come singly.

Far more difficult to diagnose, and sometimes perversely irritating, are those faults concerned with known intermittent running. The clock goes for a few days and then stops. Almost certainly if it is helped by a turn on the spring and even by advancing the hands, it will proceed to run again, but for an uncertain period. As a rule, the root cause is insufficient power—not in the clock as designed and as it should normally run, but owing to some temporary increase in load due to undesirable friction, often coupled with an escapement not in perfect order. Again, the hands are the first place to inspect, particularly the minute hand. Turn the hand, or the set-hands arbor, through several hours, noticing whether the tension on it is consistent or varying. Take the minute hand off, at a time when the clock is struggling, and see if the movement will then run with proper vigour. If the tension of the hand varies significantly or if the movement will run without it, the likelihood is that the hand has too great a tension from the collet, is rubbing the hour hand boss, that its collet is not regularly domed, or that the spring of the clutch is the wrong way round or too tight—it will be apparent if the spring is far beyond that needed to keep the centre arbor and the hand revolving together in normal running (Fig. 7). The hand pin (if any) should not control the tension but be such as to fit well when the collet is removed. The area will be localized as that of hands and motion work if the movement comes to life when they are removed; faulty motion work through a bent stud for the minute wheel is a not uncommon problem. If all this is inconclusive, the fault is probably in the train, and such troubles can be very elusive, since they may involve stiff and bent pivots, bent arbors, bent studs, bent or worn gear teeth (particularly on the barrel) and so on.

A clock which repeatedly stops a few minutes before it should strike (or chime) may have a fault in the sounding or the going side; either way, the raising of the lifting piece (or flirt), locking piece (and hammer spring if maladjusted) is too much for the train. This may be caused by bent or too stiff springs on the striking side, by incorrectly having set up the sounding train with the hammer(s) 'on the rise', or by wear in the lifting piece where it contacts the lifting pins. Alternatively, or with some of these faults, the going train may be weak for several reasons—primarily friction arising from poor gearing with worn holes and pivots or, less often, an incorrect or tired mainspring or a sluggish, poorly adjusted escapement. To establish that it is the interaction of striking and going which is causing the trouble, start the balance or pendulum by hand and lift up the hammer, locking piece, lifting piece, or all three, whilst holding the fly still; if the escapement then comes to life, the fault is evidently in the work demanded by initiating the striking, though that may be because the going side is out of condition.

Unless the clock merely runs down prematurely, the spring of an intermittently troublesome clock is usually satisfactory. Wind it up slightly and, holding the scapewheel, remove the pallets, or else, holding the contrate wheel, remove the platform escapement. Observe the free running of the train, first with the motion wheels on and then with them removed. If, without the motion wheels, the trouble seems to lie in occasional lessening of power in the train and slowing of the revolutions of scapewheel or contrate wheel the region of friction may be clear from the sound—there is not usually much difficulty in deciding where a tooth is badly bent on a wheel, since the stoppage is obviously where one of the wheels engages rigidly with a pinion. If the difficulty is not with a wheel, it may be with an arbor. With the wheels revolving fast, a bent arbor can be seen. Pivots, however, must be tested one by one (with the movement still assembled) for endshake and sideshake—that is, their movement in the holes from side to side and from end to end of the pivot. If there is insufficient endshake, such that an arbor shoulder may bind on a plate, consider whether the plates may be bent or extra pressure has been imposed at that point. This is a matter where 'feel' comes from experience but, with the spring unwound, it will not be difficult to detect anomalies in the freedom, or lack of it, for wheel arbors. Remember that too big a hole, through wear, can produce the same result as too small a hole (i.e. for a bent pivot—you are not likely to find a hole originally too small); a worn hole leads to misalignment of the wheel concerned which, in extreme cases, will not mesh properly with the engaging pinion. Note, however, that if the train is seriously slowed as it approaches the time of striking and there are no other defects (for inevitably there *is* increased load here), the fault may lie in a bent lifting and warning piece which, in bad instances, will stop both striking and going trains.

If the clock will not go and the power is demonstrably transmitted through the train, the difficulty is likely to lie in the escapement, and some adjustments to this are outlined later in this chapter. One intermittent fault should be mentioned here, however, and that is with the platforms of carriage clocks. It is always worthwhile with these clocks (when you have ensured that the contrate wheel depth screw is not too tightly set) to loosen the screws on the platforms and to try the effect of running the clock with the platform in a slightly different position. The contrate gear with the scapewheel pinion is very sensitive to pressure, both direct and at an angle, and the difference of platform position between where the movement will run well and where it will hardly struggle for more than a minute or two, is often very slight.

Many of these faults of going trains apply equally to striking and chiming trains, where evidently even transmission of power is still required, though there is more velocity once the train is under way. If the sounding trains will not run, the trouble is again, provided you can see that they are properly released, likely to be in the wheels or their pivots and holes. One fault more than any other, however, is responsible for the failure of striking and chiming trains, particularly when their power starts to run down, and this is incorrect setting up in the first place. It is obvious that the resistance of the hammer or hammers and their return springing is very considerable and it follows that the mechanism must be so put together that the hammers are off their lifting pins (or teeth on the star wheel) when the striking is stopped; if this is so, the train will have developed some momentum by the time blows are actually struck. Maladjustment here is also a contributory factor to inaccurate striking, especially in the countwheel system, for it can result in the first blow of, for instance, the eight o'clock sequence being struck after the last stroke at seven o'clock, with the result that the countwheel fails to lock properly and the sequence thereafter is permanently out of step. There should also be an idle run for the warning pin if striking is to be correct, or the train will either not start or will strike a preliminary blow at the warning. Again, with the rack system there must be an idle run for the gathering pallet or in the worst cases it can arrest the rack before it has fully fallen. It should be added that worn and enlarged holes for the pivots can cause inaccurate striking, and particularly the holes for the locking wheel and the wheel, if any, on whose arbor the countwheel is fitted. Check sideshake in these holes closely.

The other recurrent faults and adjustments of striking and chiming are mainly peculiar to the system involved and are dealt with below. The exception is the lifting or letting off mechanism common in some form to all systems. If the striking runs well usually but sometimes misses, and all is well with the train, or if it misses on some quarters or hours, the trouble is likely to be located here, and it may be associated with the hands and motion work generally. It is obvious enough, but a little awkward to check, since on many clocks the lifting pins are not visible, being at the back of the cannon pinion, or at any rate obscured partially by the hour wheel. A bent lifting piece may not be contacted or raised by the pins, or may be raised irregularly by the surface of the cannon pinion (or minute wheel) rather than only by the pins—this will cause extra striking and may bring the clock to a halt if the lifting piece jams.

A bent warning piece may also not rise far enough to come into the way of the warning pin, or may not fall far enough to release it. In the latter instance it may be so high as to butt against the plate and stop the clock completely. In the French system, the pin by which the warning piece raises the rack hook may be loose so that the rack is not fully released. The graduated pins used for some rack striking systems may be bent or incorrectly placed so that the wrong number of rack teeth may fall (or the rack itself may be bent and crooked, or not free on its stud). Half hour striking needs careful adjustment in this respect, or several blows will be struck. Badly worn lifting pins do not always give sufficient lift for the system to be let off. Then again, the minute hand may well be sufficiently well pinned on to turn with the centre arbor for timekeeping, but still not press the cannon pinion far enough back for its pins to engage with the lifting piece. Combinations of these faults are often present. Provided that the clock can be relied upon to get past twelve o'clock, such troubles may be tolerated for years, or the strike/silent lever may be set permanently to 'silent' to disguise the problem, but they are all such as the amateur can detect and attend to with little difficulty, elusive as they may be.

This discussion may give some indication of how to diagnose by locating the likely areas at fault. The remaining sections of this chapter consider some of these areas in more detail. The arrangement is alphabetical for convenience, although on some subjects reference should also be made to the Index.

*Alarms*

Alarm mechanisms as such seldom give trouble, and some of the principal sources of difficulty have already been mentioned—notably the setting of the alarm hand and the danger of setting an alarm backwards. The latter manifests itself variously in broken motion wheels or bent motion wheel studs and, more often, in an alarm collet or projecting pin, according to which is moved, which comes loose from the alarm setting arbor, with the result that the alarm performs erratically if at all. It might be thought that modern screwed setting buttons, screwing on in the direction of setting and coming off if screwed the wrong way, would avoid this trouble; but the buttons come off and are lost, and improvisations are made which cause the damage.

In all alarms it is necessary to adjust the clearance of the lever which intercepts the hammer so that it does not touch the hammer during sounding, yet positively silences the alarm when it is not set. Likewise, the silencing lever must be adjusted until, with modern movements, it not only stops the alarm when it is running, but also is released when the alarm is wound. Some adjustment may also be needed to an intermittent-sounding 'snooze' pawl; it must stop the hammer when on the high point of the ratchet, but there must also be clearance for this tooth, or the clock will stop.

A common fault particularly on alarms in carriage clocks, but also more generally, is that the alarm setting arbor is not held friction-tight and turns with the alarm wheel. Then the alarm sounds, if at all, only when it is set to the same time as the clock's hands are showing, and the alarm hand anyway travels round in company with the hour hand of the main dial. The remedy is to increase the strength of the coiled spring or spring washer on the arbor, inside the back plate, or to pack it with an additional washer, since the arbor keeps in place against the travelling alarm wheel only by virtue of its friction with the movement plates.

*Chains and Lines*

There are certain difficulties in running an old long-case clock with ropes, though it is to be preferred if that is the design. If, however, it is wished to change these to chains, it

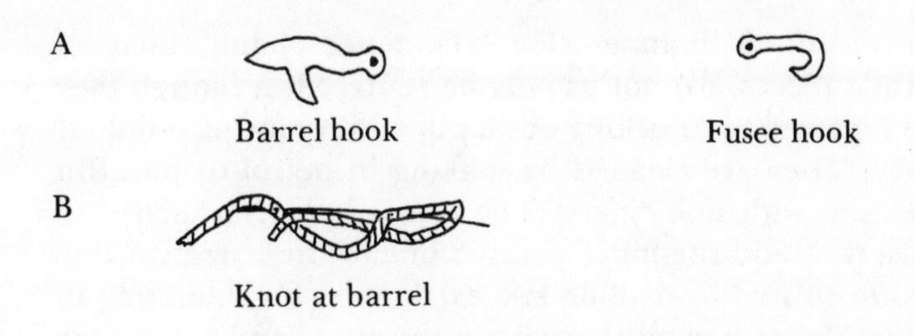


Fig. 41. Securing fusee chain (A) and line (B)

must be remembered that a chain will not run on the pulleys of a rope-driven movement, since the holding spikes are of a different shape, and it may not be possible to match the spacing. Modern chain pulleys are available with matched chains and modern clickwork, which is likely to be more reliable than the old type and is simpler to repair (Fig. 3). These pulleys usually have a groove between the spikes and this fits the vertical link of a chain whilst the spike holds the horizontal link next to it; the intention is to prevent the chain from turning on the pulley, for turning is a major cause of slipping. The old pulleys are secured by stout pins, sometimes through retaining collets, which have to be driven out, and it is possible that the arbors will have to be filed (or preferably turned on a lathe) slightly to fit the new pulleys.

Frequently slipping chains are either of the wrong size or open-linked, though occasional slipping seems unavoidable. Chains can probably be bought for old pulleys, but it is as well to provide the pulleys as a pattern, for measurement of links and pulleys is none too reliable. Increasing the counterweight, provided the driving weight remains sufficient, can help resolve the difficulty of slipping, and it should also be ensured that the movement is firm in the case.

Frayed and stranded gut lines should always be replaced, although they are of course especially dangerous in a fusee clock with powerful springs. Replacement lines may be of twisted steel wire or of synthetic material. The former have the advantage in strength, but they are difficult to secure and in time score the barrel. Gut lines should be well rubbed with oil before use. A fusee line must be secured to its barrel in the special traditional manner (Fig. 41). Fusee chains, used on high-grade pieces, cannot usually be replaced, although they may be repaired by knocking out a pin and removing a link, if necessary. They are cleaned by soaking in petrol or paraffin and treating with fine emery. The round hook is attached to the fusee post and the more pointed one to the barrel. A line can be substituted by drilling two extra holes for mounting in the barrel, by removing the post in the fusee and taking the hole right through for a knot inside the fusee.

*Chiming*

Many of the problems that arise with chiming are merely those of striking, if one can but analyse them dispassionately. There are, however, three which are more peculiar to chiming. These are matching the quarter struck with that shown by the hands, letting off the hour, and persuading the sequence of bells or gongs struck to correspond to the breaks in the running of the train.

In the full rack system the regulation of the quarters is carried out by the quarter rack, as we have seen. This is normally not on the centre arbor but on the minute wheel, which revolves anti-clockwise. Often the hour snail is mounted on a starwheel turned by a pin on the cannon pinion. It follows that when the clock is assembled, these wheels should be so placed that the cannon pinion has positively pushed the starwheel and snail round just before the fourth quarter lifting pin on the minute wheel is engaged by the lifting piece. If the hour snail is on the hour wheel in the centre, rather than offset on a starwheel, it has to be placed similarly, so as to ensure that the hour rack falls squarely on to a certain section of the snail when the fourth quarter is sounded. The hands must be fixed according to the position of the quarter snail, and then they will always indicate correctly. Where the minute hand cannot be placed exactly to match the quarters, the minute wheel has to be moved a little relative to the cannon pinion. If alignment still cannot be achieved, the hand itself must be moved in relation to its squared hole, being riveted once the correct setting is found.

The countwheel system differs, of course, in that if the chiming is somehow inadvertently released or fails to sound, the sequence will be wrong unless the chimes are corrected manually or unless a self-correcting device is incorporated. It is necessary to ensure that the fourth quarter lifting pin—in these clocks usually on the cannon pinion—is the one operating when the extra-high section of countwheel comes round, or that the lifting pin previous to it has been used to release the third quarter on the countwheel, which ends with the setting on of the correction mechanism (*see* Chapter 4). In the countwheel system the strike flirt, which acts as countwheel detent, must be in one of the countwheel notches when the train is locked by locking wheel or hoop wheel.

As has been noted, the hour striking is set off on completion of the fourth quarter striking. It can sound at no other quarter because, in the case of the rack system, only then does the quarter rack fall far enough to release the strike rack hook and rack and, in the case of the countwheel system, only the extra-high section of countwheel can raise the strike flirt high enough for it to release the strike rack. Although the hour warning piece is raised at every quarter, nothing can happen in the hour striking train until these fourth quarter arrangements come into effect and the rack is released. Then the striking is held at warning until the chiming train comes to a stop and in consequence the strike warning piece, attached to the strike flirt, or connected to the quarter rack, falls. If you are not used to chiming clocks, this has to be remembered as the fundamental fact, to which all the setting is related and the 'tune' matched; otherwise a great deal of time can be wasted trying to get the train to correspond to the 'tune', whereas it is the chiming barrel and ratio wheels which must be altered to correspond to correctly set chiming.

It has been mentioned that in proportional chiming a barrel with five sections of tune revolves twice in an hour. It follows that the first of these five sections may represent the first quarter, or the third of the three sequences sounded at the third quarter, and there is nothing to choose between them, although from a practical point of view it is obviously much easier to set the chimes just before the first quarter, when the train will be arrested rather than running. The beginning of the barrel is a place clear of all hammer pins, and it is usually followed by, for the first sequence, a downward scale, identifiable as a diagonal line of pins across the barrel. When this has been found, the barrel is turned, with a ratio wheel removed, until all the hammers are in this clear space on the barrel. Then replace the ratio wheel and check the chiming round, ensuring that all the hammers are clear of barrel pins whenever the chiming is locked. This is a close adjustment and it must be made correctly, or the train may not be able to start when released and the sequence will be distorted.

It should be added that there are books containing a fair selection of common chiming sequences and these may be helpful if you are faced with an unfamiliar one. The tunes of the eighteenth century and earlier are, however, often strange by modern standards and not recorded. Then you have to settle for what sounds acceptable—particularly as the ratio wheels may not be readily accessible for adjustment.

*Countwheel Mechanism*

Countwheel striking and chiming depend on the coincidence of the countwheel detent's falling into a countwheel slot and of the locking piece's falling into the notch of the hoop wheel (or meeting the pin on the locking wheel). Countwheel detent and locking piece are either made in one piece or are two parts fixed to the same arbor. The countwheel does not normally perform the locking and if it does so, through misplacing of the detent, the train may well fail to be let off owing to the relatively large power behind the countwheel and its friction with the detent. Thus at the end of striking—particularly half hour striking when present—the detent must not fall hard up against the next raised section of the countwheel, and neither must it be so far back round the wheel that an extra single blow can be struck. With half-hour striking, the detent must fall near the beginning of the slot, or the half and following hour will be struck together.

Countwheels are not always laid out in minutely accurate proportions. They may have round or square arbors, and their best position has to be found by experiment with the train locked and the detent tried in various countwheel slots. As a rule, their position is easily changed, since they are pinned or screwed to their arbors. It is unlikely that a countwheel, if original, never worked reliably, and these wheels are subject to little wear. Therefore the temptation to alter the size of slots and raised sections must be resisted, at least until absolutely all else has been tried. A final check worth making is for an enlarged hole to the countwheel arbor, which is under considerable power. A worn hole here can cause most perplexing performance.

In some systems a hoop wheel, consisting of a cam with a depression, is fitted as well as a locking wheel. Its purpose is to raise the detent up and down until a slot is found. Most French countwheels, however, with locking wheels, have sloped exits, the detent sliding up and down along the countwheel itself. The setting of the hoop wheel must of course coincide with that of the locking wheel if this subsidiary hoop wheel is fitted.

A common trouble with older clocks arises in the action of the locking piece, which may drop into the hoop wheel notch or meet the locking pin only to bounce free again so that the train continues to run. Such clocks may also start when they are being wound, since the winding is inclined to jolt the barrel and the countwheel arbor. Correction is a matter of adjustment to the angle of the locking piece tip and, if necessary, of strengthening its spring. The locking piece should always drop to the bottom of the hoop wheel notch and the countwheel detent to the bottom of the countwheel slots, but it must be ensured that the detent does not hit the bottom *before* the locking piece, or the train may not lock safely. Sometimes these pieces may have to be bent. In modern clocks, and in countwheel chiming, the locking piece is often mounted with a grubscrew and is readily adjustable.

A replacement countwheel can be constructed from a circle divided into 78 (hours) or 90 (hours and half-hours) divisions, the size and depth of notches being clear from the position of the arbor and fall of the countwheel detent. It is also possible to mount a disc on the arbor and, as the clock is struck round, to mark where the detent falls and where the train should stop, and then to cut the countwheel from the marked pattern.

*Cuckoo Clocks*

Although there is nothing very novel in the principles of most cuckoo clocks, the profusion of their levers and the lack of visibility of their mechanism because of their solid wooden plates can make them a little daunting if they are among the first clocks one has to repair.

These clocks are usually made of a hardwood frame and plates, with bushes to house pivots. They may be weight or spring-driven, sometimes transmitting the power externally from a pinion fixed to an extended arbor of the pulley wheel or intermediate wheel (when, as in the old single-handed thirty hour clock, there is no centre arbor in the usual sense). They normally employ countwheel striking. The bird is operated by a vertical cranked wire connected to the countwheel detent and locking piece, and is so constructed that (if with movable wings) its wings are opened when its tail is raised by the opening of one of the sound bellows. The doors are hooked to the base of the bird, which opens them as it is pushed forward by the vertical wire.

The lifting piece and warning piece, which is of bent wire, are on the same arbor, and the locking and warning may both take place on the same wheel and pin. There are three arbor tails all working on the same pinwheel, the first for the gong hammer, and the other two linked to wires going to the top of the sounding pipes, where they raise the small bellows, whose falling by their own weight produces the cuckoo call. It is necessary to take care with the linkages of these wires and also with their lengths, especially if they are moved to fit new bellows; too short a wire will produce a poor note and too long a one will stop the striking train because the pinwheel will not be able to pass the hammer tail. Where locking is on the warning pin it must be ensured that, when the train is locked, the locking piece is also down in the notch of the small cam or hoop wheel which moves the detent up and down in search of a slot on the countwheel.

Most ancillary parts for these clocks can be obtained new, as can hands and sets of plastic figures for applied dials. It is possible to make pipes from cigar-box wood and bellows from kid-glove leather, using old ones as patterns. Sticking patches over bellows, with whatever materials, is rarely a satisfactory solution. If only a bush is worn, it is easily replaced. If the wooden wall of the hole for the bush is defective it has to be plugged and well glued so that it can be drilled for a new bush.

*Electric Clocks*

Electric clocks are of such variety that it is not possible to cover them in detail here. It should, however, be said that in general the motive power involved is very small and the mechanism simple. The most modern movements are mass-produced with extreme economy in mind and while it may be a matter of interest to repair them, especially if worn, it is often not economic, since new replacement movements, which are compatible if not identical, are available.

By far the most common fault in older battery clocks, such as the Bulle, is breakdown of the insulation of the connecting leads. Wires are often clamped between metal parts, or run down inside metal columns, to make them unobtrusive or invisible, and in course of time the insulation deteriorates. Particularly is this so of insulated contact pins and spring contacts. Once the principle of working has been discovered, it is simple to test the circuit with a flashbulb and battery at a time when there should be no continuity, and then to set about finding where the breakdown has occurred.

The other point of weakness is at the contacts themselves. These must obviously be free of oil and dirt and should be cleaned with a fine file or emery as a matter of course. Modern contacts are sometimes sealed into a vacuum tube and operated by the approach of a magnet, so that this problem does not then arise.

Owners of older battery clocks are not always aware that large-capacity batteries are needed to keep them running for a period; a small torch cell will not suffice as a replacement. The Bulle clock must have its battery connected the correct way round for the clock to run, but actual damage will not be caused if it should be reversed.

The modern re-wind and magnetic balance models are not subject to the same weaknesses. Here the adjustments required are usually mechanical, either to the position of the worm driving the motor, to the freedom of a contact arm, or to the spiral driving sections on the balance staff—making this drive the 'scapewheel' reliably may involve slightly moving the balance by means of its bearing screws or bending the metal drive slips so that 'scapewheel' teeth are not fouled. The re-wind type normally uses a fine pin-pallet escapement and this must of course be in good order in view of the very small power delivered by the mainspring. In the version using an electro-magnet rather than a motor, the springing of the armature has to be adjusted so that the armature jumps clear of the contact once the spring is rewound. Some balances impulsed by an electro-magnet or coil have to be set up 'out of beat', else when the current is switched on they will not start. The coil or coils must be midway between the two balance wheel rims. They are delicate and must be handled with great care. With transistorized circuits the battery must be connected in the correct polarity or damage can be caused. It is an elementary step with all failed electric clocks to check continuity of coil windings and for short circuits due to broken down insulation.

Synchronous clocks are very susceptible to stiffness in pivots, rubbing of the hands on the dial, and other wastage of their small power. A synchronous motor must run at the speed dictated by the mains frequency—under excessive load, it will stop rather than run slow. In the event of stoppage the first check should of course be made to the connecting lead, plug, fuse if fitted, and insulation generally. Thereafter, the trouble in the motor may be caused by an interrupted field coil, a seized rotor, or obstruction brought about by a bent starting lever or a stiff idler pawl in the self-starting motor. The pulsations in the field coil can be felt with a screwdriver and, if it is being magnetized, the fault must lie in the rotor or train. If pulsations are not felt, and the connections are in order, disconnect the clock from the mains supply and check the continuity of the coil with a dry battery or low voltage supply. Check also for a breakdown of insulation between coil and core. Unsatisfactory results for either indicate replacement or rewinding of the coil. The bearings of some types of rotor run dry after a period and should be lubricated with two or three drops of machine oil into the felt pads. The idler pawl working on self-starting rotors of the 'spider' type must be completely free in its movement.

The majority of electric clocks are now fitted with centre seconds hands, partly to indicate that they are running, or with an arbor for such a hand which can be blanked off if not required. Any friction of the seconds hand on the pipe of the minute hand if carelessly pressed in beside the arbor rather than on to it, or if the movement is placed face-downwards during examination, is liable to stop the clock.

*Escapements*

These paragraphs enlarge on general descriptions of escapements and their working given in Chapter 3.

The essential requirements of any escapement are that it should lock positively and that it should give the balance or pendulum sufficient impulse. In practice, this necessitates an appropriate depth of the pallets into the scapewheel, correct angles on the pallets, and a reduction of free 'drop' to the minimum necessary. The locking and the drop (the space between a pallet and a tooth at the moment when a tooth is released by the other pallet) should be the same for each pallet and each tooth of the scapewheel. If they are not, the wheel is out of true or a tooth requires stretching or reducing; it is undesirable, for instance, for an escapement to run shallow because it has to do so in order for the pallets to clear one tooth minutely longer than the others.

The means of adjustment naturally varies according to the escapement. In older clocks there is often no adjusting facility. The pallets have to be raised or lowered by tapping the suspension cock up or down, and opened or closed by tapping the belly of the pallets, softened by heat, over a vice with a hammer. Sometimes steady pins have to be moved and the cock mounting holes enlarged. On more modern clocks the suspension cock is usually mounted in slots rather than holes and is easily moved, and the front pallet hole may be mounted in an eccentric hole in a friction-tight block, which can be turned for adjustment. Strip pallets can be opened or closed after softening. Many French and Vienna Regulator movements have the pallet arms running parallel before they joined the arbor, and separated by their springiness against stop screws; moving these screws opens or closes the pallets.

The depth of cylinder escapements is controlled by the 'chariot' in which the lower balance jewel (and balance cock) is set, but altering the depth of lever escapements involves moving the pallet stones or pins, or bending the pallet arms, neither of which is to be undertaken lightly since the required movements are so small. Jewelled pallets are normally mounted in shellac, whether on pendulum or lever escapements, and can be moved if gently warmed. If one pallet is moved, the other has normally to be oppositely adjusted to the same extent. Pin pallets are brittle and cannot be bent. They are easily replaced by driving out with a punch and fitting hard steel pins of suitable size. The pallets of Four Hundred Day Clocks and Vienna Regulators are, as already noted, usually dropped into place in the arms and clamped by screws; they are easily reversed, but are not interchangeable in that the two members of a pair are quite different. Brocot escapement pallets, of steel or jewel, can be replaced with steel rod filed semi-circular and driven in or mounted in warm shellac. The flat surfaces must be radial to the scapewheel and the rod should be just slightly less in diameter than the distance between two wheel teeth. The draw of lever escapements is adjusted by moving the pallet pieces so that the wheel teeth push them firmly into banking. In the case of pin-pallet escapements it is produced by the undercut faces of the wheel teeth and must be increased, if necessary, by fine filing of the teeth. Banking for all lever escapements must be adjusted by bending the pins or modifying the sunken 'walls', so that banking occurs late enough for the balance to revolve quite free of the sideways lever, but not so late that the impulse pin cannot cleanly align with the lever fork on the return swing.

A shallow escapement trips. It keeps poor time because feeble locking occurs, partly on the impulse face of the pallet, and it is liable to stop because the full face is then not available for impulse. Deepening an escapement will increase the arc of the oscillator, but the pallets of too deep an escapement are liable to butt on the back of the preceding teeth and, at the worst, the scapewheel will fail to be unlocked—entirely, or regularly on one tooth if the wheel is faulty. If there is insufficient drop, a pallet will land on the point of a wheel tooth rather than on the far side of it—again, the trouble may be caused by a bent tooth—and the action will jam or the scapewheel will not turn. If there is excessive drop there will be poor impulse and weak locking, and increasing the depth will not cure the trouble. Such faults may be irregular owing to wear on the pallets and pivots, endshake, and small unevenness in the scapewheel. The latter can be filed out on an individual tooth, but is best done by 'topping', revolving the wheel, supported by an accurate chuck, fast but very lightly against an abrasive surface, and then filing the shortened teeth to shape. Be careful before being tempted to file or alter the radial or undercut faces of scapewheel teeth, for if this accurate division of the wheel is altered there is very little hope that the escapement will run correctly again. Adjustment should always be made to the sloped or curved backs of teeth. It is unwise to make any adjustment to the teeth of a clubtooth lever escapement scapewheel at all, for these wheels are modern, made from steel and wear little, and the profile of the back 'club' is essential to working.

Wear, resulting in troughs in the pallets and causing shallowness and excessive drop, is the most common trouble with anchor-type escapements. The wear is best stoned out with a carborundum stone. After softening the pallets by heating to red heat they can be closed slightly if necessary to increase the depth. It is also possible to resurface pallets with soldered strips of steel mainspring. Alternatively, it may be possible to move pallets or scapewheel along their arbors so that the unworn surfaces can be engaged. Brass collets are soldered or driven on and not hard to move with a hollow punch, but proceed carefully in tapping directly mounted steel pallets along their arbor, for they are very brittle, especially those in French clocks. After resurfacing pallet faces, harden by heating and plunging into cold water, and then polish highly with graded emery. The final polishing should be in the direction of the pallet action. It need hardly be said that the pallet holes, from which the crutch also hangs, are subject to wear, and the escapement cannot function reliably if there is excessive shake here, for impulse will be lost and mis-locking is likely. Excessive endshake must be taken up by bushing the holes proud (*see below*), or the pallets, unless perfectly regular, will present a varying surface to the scapewheel teeth.

The action of lever, including pin-pallet, escapements has to be carefully observed. Take the balance out and check the wheel round tooth by tooth, noting the locking and that the lever, under the action of draw, swings sharply over to the banking pins (or to the bottom of the wheel teeth if the pin-pallet escapement has no banking pins). Then, with the balance central, check that the lever is in beat and has to move the same amount in either direction to release a tooth. Rectify any inequality by careful bending of the lever, unless it is of steel, in which case set in beat by moving the hairspring on the balance only. Finally, check the action of the lever fork, that it is fully entered by the impulse pin and that the horns do not scrape on the staff. Ensure also that the guardpin, or point of the lever, does not rub the roller but only permits the lever to move when it is in the passing bay of the roller or balance staff. The lever can be lengthened by tapping in the middle, or shortened by bending a section down a little with pliers, making another bend to restore the lever to the level.

Although it may be an overworked practice, there is nevertheless often good reason to replace a cylinder platform, typically in a carriage clock by a new lever platform. The repair of a broken cylinder staff will cost only marginally less than a new platform. For many amateurs the repair must be contracted out, and even then the escapement will be worn and will suffer from the defects (particularly the small arc) of its type. New platforms come in various sizes and naturally the size nearest to that of the existing platform is chosen. The only critical point in matching is the scapewheel pinion, for the variation in time-keeping caused by using a pinion with the wrong number of leaves is too great to be corrected by modifying the balance spring. If the original platform is missing the required pinion-size can be calculated by working out the ratios needed between the revolutions of the scapewheel (dependent on the vibration of the balance, a figure which the supplier will provide) and of the centre arbor, revolving once in an hour. Once the new platform is to hand, it has to be stripped of the balance and escapement so that the plate can be drilled for mounting. The holes may well not be in the same place as those of the superseded platform, so it is as well to drop the new escapement into place for a test run, and to mark the plate according to the best position for mounting. Drill the holes on slightly the large side so that there is a little scope for adjustment when it is fitted. The mounting screws are large-headed and often provided with washers to cover such holes.

*Gearwheels*

The principle of gear ratios was mentioned in Chapter 1 and it is not proposed to go further into the subject here. Suffice it to say that these ratios are found by multiplying together all the teeth of affected wheels, doing the same for all the pinions, and finding the ratio of these two products to each other. The ratio required is of course that of revolutions of different wheels in the same period, and the sums are done using known factors—vibrations of a pendulum or balance, blows of a hammer—as points of reference. Care has to be taken not to confuse the issue by mixing wheels concerned only with duration of running (for example, the barrel and intermediate wheels) with wheels and pinions concerned with the measurement of time, in the sense that they come between escapement and hands. The centre wheel is concerned with time, but its pinion is concerned with duration of going.

A broken wheel can be repaired when only one or two teeth are broken, the procedure being to flatten the slot and to dovetail in a new piece of brass, soldering it in place and shaping it when fitted. The slot for the tooth is formed so as to present as large an area of contact with the wheel as possible (Fig. 42). The wheel should then be 'topped' as mentioned previously. Bent teeth can sometimes be straightened or filed into usable shape but there is always the risk that they will break or slip under load. Broken barrel teeth can be replaced by levelling the space and drilling it, to at least a tooth's depth, then driving in one or two steel pins, to be pointed to the profile of a tooth's tip. It is worth noting that barrel bases tend to be soldered to barrels; they can be removed for repair or replacement. Most bigger jobs have to be sent out as the facilities for accurate wheel cutting are not generally available to the amateur. However, where wheel teeth are badly worn, this is usually on one (the driving) side only, and it is sometimes possible to drive the wheel from its collet or pinion seating, reverse it and rivet it back on with a hollow punch.

A bent arbor may be straightened by many taps of the hammer at the apex of the bend, but sometimes of course a new arbor is the only solution and you will require a new pinion cut at the same time. Arbors in French clocks are

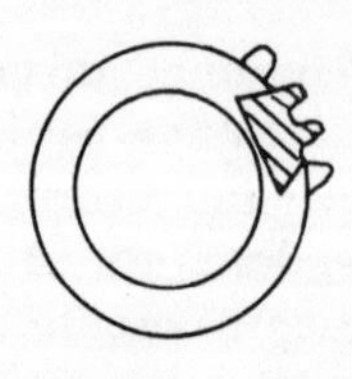
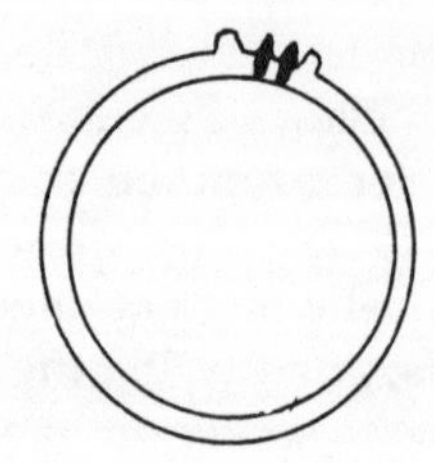

Fig. 42. Repairing wheel and barrel teeth

extremely hard and brittle and require softening first.

There is no satisfactory repair for a badly worn or broken pinion of the leaved type and a new one will have to be cut. Lantern pinions, however, which consist of brass collets with steel pins between them, can usually be repaired. As a rule, the pins (trundles) are pushed through one hole and the end riveted over, whilst they lodge in the blind hole of the other collet (shroud). A broken trundle can be pushed out, the riveting being scraped back if necessary, and a new steel pin substituted—the new pin should be of exactly the same gauge. The shrouds must be straight and should be sweated to the arbor with soft solder if one is loose enough to turn in relation to the other.

*Holes and Pivots*

The holes and pivots are of such basic importance to a clock that it is surprising how often they are neglected. A worn and crooked hole means a crooked arbor, imperfect gearing, unreliable action and, eventually, the chance of serious breakage from the release of the power of weight or mainspring. Holes and freedom of pivots in them (shake) are, however, judged by experienced feel rather than by measurement and inevitably circumstances arise when attending to a worn hole is put off and left to the next repairer.

A pivot must have satisfactory endshake and sideshake—the arbor must be able to move between the plates, so that the plates do not grip the pivot shoulder, and the pivot must have freedom to move up and down, side to side, in the hole. The endshake may be unsatisfactory in the first place, or arise from distortion of the plates, whilst excessive sideshake is usually caused by wear.

Excess endshake may be taken up by bushing the hole, as described below, and leaving the bush slightly proud. It is unwise to try to punch the plate in, as this causes distortion and may affect other holes. The best cure is of course a new arbor properly pivoted to the right length. Lack of endshake may be cured by bushing with the bush recessed, or by lightly punching the plate back on a stake at this hole, after which the hole will need to be opened with a broach.

Too large holes are almost always crooked holes, caused by the pressure of the driving wheel against the driven wheels. They must, therefore, be moved (drawn) straight and then bushed. Drawing is done with a round file, increasing the oval on the opposite side so that the centre is once again where, so far as can be judged, it should be. The hole is then rounded and enlarged as necessary, from behind, with a broach and a brass bush is tapped in exactly upright, again working from the inside of the plate outwards—this imparts a slight taper so that the bush cannot slide out. The bush should be a close fit, but not so tight that it is distorted when it is knocked into place. It may be left proud or recessed to adjust endshake, and should ideally be riveted over on the outside, the recess for oil (oil-sink) being countersunk into the new metal. Once the bush is firm, it will probably need light broaching before the pivot will run in it. Then the particular wheel and its neighbours must be tried between the plates to ensure that they run freely and smoothly.

It is possible to tighten holes by punching dots round them or by using a hollow punch to make a ring, so pushing in the edge of the hole. Neither method recentres the hole nor produces a proper wall to it. They should be reserved for the cheapest clocks repaired in haste, if they are used at all.

If a good chuck is available, pivots should be burnished and polished in it. Otherwise you will have to make do by hand with a virtually toothless old file. Use a hardwood block with a notch in which to rest the pivot, and twist the arbor to and fro against the movement of your burnisher charged with oil. With such equipment fine pivots are best left untouched, save for the necessary removal of rust and debris. A severely worn pivot must be replaced, and its hole must be bushed since it will usually have been gouged. It is within the home repairer's province to drill the arbor of a long-case or large bracket clock and to tap in a steel pin to act as the new pivot. The pivots of French clocks, alarm clocks, platform escapements and so on will generally have to be sent out, when either they will be similarly pivoted with the accurate drilling facility of a lathe or a whole new arbor will be turned. Bent pivots can sometimes be straightened after softening in red heat, and it is most easily done gradually by means of small movements of a hollow punch placed over the pivot, though it can also be carried out with tweezers or pliers. A straightened pivot must be tested between neighbouring wheels in the plates, for appearances can be deceptive outside the movement.

*Mainsprings and Clickwork*

Mainsprings broken at the inner end cannot be repaired, those broken in the middle are best replaced (for repairs here do not last), whilst those broken at the outer end are not only the most numerous but also the most amenable to good repairs.

The method of repair of the outer end naturally depends on how the spring is fixed. Open springs with no barrel have normally only to be softened and bent to the shape required for their securing posts, though holes may have to be drilled to rivet a loop. For other springs a hole has generally to be drilled or punched to receive the barrel hook. This hole must be central and is best kept round, or the corners are apt to break away. Soften the metal by heating to bright red and allowing to cool slowly, and then drill a small hole on the site of a shallow punch mark, broaching and filing the hole to the size required. It is possible to make the initial hole by filing through the dimple left by a deep punch mark, but such punching may weaken the spring. The tip in front of the hole is best left slightly shorter than the width of the spring. Too long a tip will make the spring hard to wind into the barrel and will lead to a crack at the hole, while too short a tip may break away. A new barrel hook can be made from steel rod or a screw with its head removed. The hole in the barrel can be very thinly tapped and the new hook forced into the barrel, ideally from the inside, before being sawn off to length and riveted over on the outside.

Attention should be given to the fastening of the inner end of the spring to the arbor hook. Often the tip of the spring, or its last coil, has to be bent to ensure firm hooking, and sometimes the hook has to be filed to improve it as a catch. This must be checked before the clock is assembled, for it is a source of real annoyance if the spring becomes detached later. In barrels, make sure that the thickened dome which supports the arbor does not obstruct the spring and that the barrel can move freely on its arbor when the spring is inside. An insecure top to a going barrel is a great danger since if it comes off the barrel may run crooked and strip gear teeth. The groove for the top can be improved with a file or graver, though it strictly requires turning on a lathe. Tops can be stretched by appropriate hammering towards the edges underneath. They can be removed by tapping the reverse end of the spring arbor but it is preferable to prise out with a screwdriver. If there is a dot on the barrel wall, the slot on the cover should be placed over it.

Ideally a mainspring winder should be used. However, although there is always a risk of distortion, most springs can taken in and out of barrels by hand, though it is wise to wear leather gloves and to extract a large spring under a blanket or sack, because of the danger of a flying barrel. In principle, all springs should be extracted, cleaned with an oily rag, and replaced with oil or thin grease. However, without a mainspring winder, a powerful spring is best left undisturbed although some oil can still be given to it in position.

The repair of clickwork is usually simple enough, but it must not be ignored, for the strain here is great and the consequences of a breakdown can be serious. The clicks of old thirty-hour clocks were springy steel circles riveted to the pulleys and with a raised part to catch the spokes of the wheel which they drove. These circles rust, split and come loose, and, whilst new ones can be filed up, there is sometimes a case for replacing with a modern pulley and click (Fig. 3), which consists of a sprung brass trigger also working on the wheel crossings. Most other clicks are steel pawls held in position by springs, though the shapes vary very considerably. The points to check are full engagement of click and clickwheel (both of which can be badly affected by wear), the riveting or screwing of the click (where a stripped thread may necessitate drilling and tapping for a new screw), and the adequacy of the spring. Some springs are soft and will not hold their clicks positively against the wheels. They should be hardened by heating and plunging into cold water, then heating again and immersing into oil to temper them. Wire springs, common in the click-work of clocks with open mainsprings, are often improved by bending, and are easily made if necessary. Springs must not be able to move out of place during winding. They may have no steady pins and move dangerously, and then it is a fairly simple matter to drill spring and plate for a steel pin. Modern click-springs are often available new, or in forms which can be adjusted—which is fortunate, for spring steel is not easy to work and these shapes cover a wide range. Clickwheels, where grossly worn, cannot be made effective merely by filing out wear or increasing the tension of the clickspring. Often they can be interchanged between clocks, but in any case they are not too difficult to cut and file from a thick brass blank or old wheel.

*Rack Striking*

The number of strokes in rack striking is controlled by the distance the rack falls, which in turn is governed by how far the rack tail can fall on to the snail. The angle of the rack tail to the rack can often be varied, the join being friction-tight, so that the teeth of the rack are cleanly engaged by the gathering pallet and so that, for example, the rack falls exactly one tooth when the rack tail strikes the highest section of the snail. A jamming or slipping rack is more likely to be caused by incorrect adjustment here than by a faulty rack hook, misplaced gathering pallet, or worn rack; but these are also points to consider, the latter especially if the rack always sticks or slips at the same tooth. To alter a rack is a drastic step, but if the rack is certainly the cause of the trouble a tooth can be stretched with hammer and punch and its profile corrected by filing. The racks of some modern clocks have no separate rack tail, but rather a sprung pin acting that part—here adjustment is neither possible nor likely to be needed.

The action of rack hook and locking piece should be carefully checked and the strength of the spring adjusted if necessary. The commonest fault is for the locking piece to graze the locking pin when the rack is being gathered, and this is adjusted by bending the locking piece outwards slightly against the rack hook, or by moving the locking pin inwards. In the old English system the rack pin, against which the pallet tail stops, locking the train, can come loose and cause mis-locking. The same applies to the occasional French movement using a similar system with the pallet tail engaging a pin on an extended rack hook. Again, a bent warning piece which, when released, sticks and does not fully clear the warning pin, can give trouble. It is possible to stretch and reshape a worn French rack hook, but it is not easy, and one should be absolutely sure that the hook is causing difficulty before one tampers with the profile of the double rack hook used in quarter-striking carriage clocks.

The position of the snail on the hour wheel should be so adjusted that the rack tail falls into the middle of each segment at the hour, or, if the steps are not marked on the snail, so that all the hours are reliably struck. Snails mounted on star wheels with spring jumpers should be so adjusted that the hour strike is let off by the lifting pin just after the snail has moved, and the jumper spring should be bent so as to give a clean action to the change-over. Care has to be taken that only the pin on top of the cannon pinion can advance the snail—if either the cannon pinion or hour wheel rubs the snail, bizarre striking will result.

Many, perhaps most, rack tails have sprung tips or are made of thin springy metal with the tip bent outwards. The purpose of this is of course to prevent the clock's stopping if the rack for any reason is not raised at twelve o'clock (Fig. 20). The springing should be adjusted as well as possible, but it is not always successful in its aim, and the overriding necessity is to ensure that the rack tail lands positively on a segment of the snail. A missing snail is relatively easily made from thick brass. The largest diameter will be found from where the rack tail falls when one tooth of the rack is released, and subsequent segments are located by moving the hands round hour by hour and marking on the brass blank, or cardboard pattern, where the rack tail must fall to secure the correct number of strokes. These points are then taken as the circumferences of circles, and the wheel divided into twelve for the steep edges to be marked and the steps to be cut. Making a rack without a pattern is more complicated and perhaps better done professionally.

The success of the flirt system depends on the freedom of the pivoted arm and the adjustment of the return spring so that it positively releases the rack hook but does not impede the going train. It should be noted that the gathering pallet of this mechanism is double; the large tooth at the rear releases the flirt from the rack hook pin as soon as the gathering pallet revolves.

Ordinary 'ting-tang' quarter-striking clocks and quarter-striking carriage clocks both employ a lever moved hourly by the minute wheel. This is in the one case to silence the high-noted gong hammer and in the other to prevent the hour rack from falling during quarter-striking. The minute wheel must be so placed that its pin moves these levers between the third and fourth quarters. (Very often there are dots punched on the motion wheels and flats on pinion leaves; if the flattened leaf is placed opposite the dot on the wheel, the motion work will usually be found to be correctly positioned for its various functions. The dots are not always original, however, and sometimes pins have been moved so that they are unreliable as an indication.)

Repeater mechanisms on carriage clocks can be troublesome, though simple enough in concept. A forked lever is used, pivoted on a stud offset from the centre. One arm intercepts the fly when the button is pushed, but clears a slot in the fly when the button is released—so ensuring that the rack falls fully before the clock begins to strike. The other, long, arm knocks out the rack hook and, in a quarter-striking clock, usually also the lever which prevents the hour rack from falling at quarters. There should be very little sideshake on the repeating lever's stud, which must be firmly screwed home, and the lever's spring should act strongly. Button and lever should, however, act quite freely. The lever terminates in an angle piece engaged by the button, and the height and angle of this have to be adjusted. There is often little clearance in the clock case and the button can slip on to the wrong side of the angled piece and fail to press it down correctly when operated. Sometimes it helps to solder a small piece of brass to the lower end of the button in the case, enlarging the area of contact with the angled piece on the movement.

*Chapter Nine*

# ASSEMBLY

Assembly is primarily a matter of replacing the cleaned and repaired parts in the reverse order to that in which they were removed, and if you have followed a systematic procedure in dismantling the clock, it presents no great problem. If there are fixed pillars on one plate it is usual to take this as a base and to place the parts in their holes in this plate, then positioning the pivots beneath the under surface of the other plate until all go home, taking great care that undue force or pressure is avoided at all times to prevent the breaking of pivots. The extended arbors are first to be located, then the barrel arbors, finally finishing with the fine pivots at the end of the trains (*see below*). Finally, the motion work is fitted and the going and striking are tried by hand before the clickwork is replaced and the movement wound. If there is not a plate with fixed pillars, it is generally best to secure the pillars to the back plate and then to proceed similarly. Through either plate there will be long arbors protruding and these can hamper assembly, as well as being—in the case of a centre seconds arbor—fragile. Hence it is a good practice to mount the first plate over an open box or jar so that it rests flat on top with the arbors out of the way inside.

Two matters commonly require attention early in the putting together. The first is a fusee, if fitted. If this has a chain there is no problem and the chain is connected later, but if it has a line or wire this must be connected to barrel and fusee before assembly and so arranged that it does not become twisted round a pillar or arbor when the movement is together. This will mean ascertaining the direction of winding, for a stiff wire coming through on the wrong side can be difficult to move after assembly.

The other question is the proper setting up of the striking (Fig. 43). There are golden rules here which must be followed if the striking (or chiming) is to be correct and, though wheel positions may alter during assembly and have to be changed

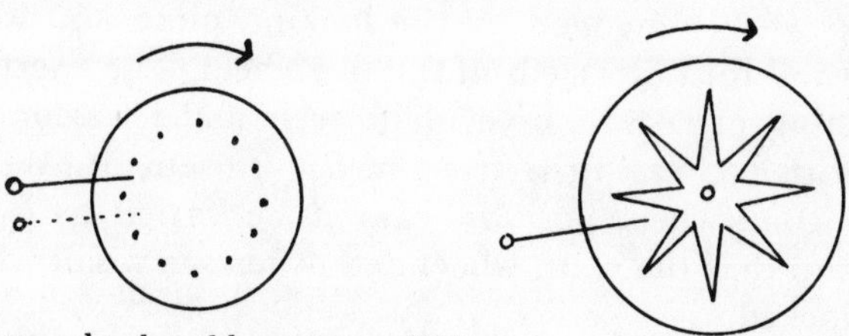

Pin or star wheel and hammer tail(s)

Locking or hoop wheel and locking piece

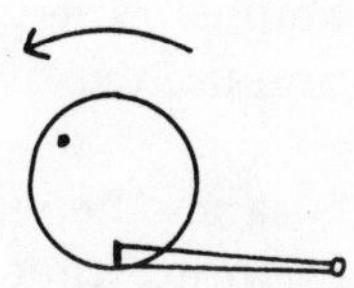

Warning pin and warning piece

Fig. 43. Setting up striking with the train locked

with one plate lifted, it is worth trying to set the train correctly at this stage. The rules are that the hammer tail must be between pins on the pinwheel (or, in practice, have just come off a pin) when the train is locked, and that the warning wheel must have at least half a turn to run before it will contact the raised warning piece. The same, of course, applies to the hammer tail and starwheel teeth when a starwheel is used instead of a pin wheel. Assemble the parts with the locking pin (or hoop wheel) engaged by the locking piece and with the hammer tail (or tails) clear of the pin wheel. The aperture for the warning piece will be evident even if the warning piece itself is fitted later to the front of the movement. Where an internal countwheel position cannot be varied by a screw, ensure also that the countwheel detent rests in a slot when the train is locked.

Fitting the top pivots into their holes can be very difficult, and little counsel can be offered. Use tweezers and work in a good light, keeping a gentle pressure of the hand on the upper plate. Take great care, for pivots are easily broken. If, as in many French clocks, some arbors pass through sizable holes and are pivoted in separate cocks, fit these cocks—which facilitate adjustment of the striking—later after partial assembly of the movement. Sometimes pivots are graduated in length from bottom to top of the movement and this can be helpful. In any case, it is usual to locate the barrels and large wheel arbors first and to pin the bottom end of the movement loosely whilst the finer pivots at the top are fitted in. According to the grade of the movement, newly cleaned plates are best pinned with straight new pins cut to even length rather than with the old ones cleaned up. Handle the cleaned plates with tissue paper and use tweezers to handle small parts to avoid soiling.

When the trains are safely between plates, revolve the striking train by hand to ensure that it is correctly set up as described above—it is best to slide the hammer on for this purpose. If now, when the train locks, the hammer tail is resting on a pin of the pinwheel there is no choice but to loosen the top plate and to keep adjusting the positions of the wheels until they are correctly positioned. In most cases, of course, chime barrels are mounted outside the plates and their position in relation to the hammers and locked train can be adjusted later, but it is still necessary to set up the locking and warning sequence correctly. Then, at this stage, it is as well to oil all front pivot holes, since they may be made inaccessible by the motion wheels and striking work when assembly is complete.

Many of the points arising in connection with assembling motion work and hands have already been mentioned in previous chapters. The important point to be remembered is the purpose of each wheel and pin on it. The motion work of time-pieces (i.e. non-striking clocks) causes little trouble as the wheels merely revolve and activate no further mechanism. The exception is an hour wheel or intermediate wheel with a pin to drive a calendar dial. If the hands are fixed rather than adjustable, this pin will have to be placed correctly during assembly—normally so as to operate at midnight. With striking and chiming, however, the lifting pins, if on the minute wheel, and with a fixed hour hand, must be placed so that the release occurs at the correct time. With quarter-striking arrangements, the lever silencing the one gong hammer, or preventing the hour rack from falling, must be moved just before the hour and the minute wheel fitted accordingly (not forgetting—as is easily done—that it revolves anti-clockwise). If apparently original, a punched dot on a wheel should be placed so that it is nearest the flattened leaf of the pinion with which it engages. An alarm index requires only the fitting of the alarm hand correctly to correspond with the time indicated by the clock when the alarm train is released.

Clocks with open springs normally have self-contained click-work on the spring arbor and wheel, and the weight-driven clocks likewise, but clocks with springs in barrels now require to have the clickwork fitted. In two cases, both of stop-work and of a fusee, there is the complication of 'setting up'—to prevent the weakest part of the spring from acting on the train and to ensure that a fusee line is taut even when the spring is run down. With stop-work, wind the arbor nearly a turn as you hold the barrel and fit the stop wheel in such a way that the arbor can turn only if wound up further. With clocks fitted with a fusee, place the ratchet wheel on the mainspring arbor and fit the associated click but leave it disengaged. Fit the line or chain, and carefully and regularly wind all of it on to the barrel. Then set up the spring nearly a turn, against the tension of the line, and push the click in against the click wheel, screwing it tightly. In future these pieces will of course remain permanently fixed and the clock will be wound by turning the fusee winding square, the winding clickwork being contained in the base of the fusee.

When you have partially wound the clock, you should now find that the replaced pendulum crutch vibrates, once the correct depth has been found, or the balance wheel escapement comes to life, and that striking is in principle correct. Complete the oiling of pivot holes and other engaging surfaces, but put no oil on the wheels or pinions of the train, which should run dry. Add dial and hands, making sure that they coincide with striking and alarm (*see* Chapter 5). There remains the question of adjusting the clock to run evenly in its case and preferably in the position which it will occupy.

Clocks will not run satisfactorily, whether mechanical or electrical (and involving a balance or pendulum), if the work required of the oscillator is demanded near one or other extreme of the vibration, and the impulse is most effective if given near the point of rest. In other words, when the pallets unlock a tooth and receive impulse, they should do so on each side at the same distance from the centre, or so that the balance or pendulum is equidistant on each occasion from its point of rest. The condition, that of being 'in beat', is brought about by slightly bending the pendulum's crutch, or turning any adjusting device, in relation to the pallets. The clock is most nearly in beat when its 'tic' is regular rather than sounding like a jog-trot. Adjusting the crutch is of course the proper way of doing what the owner tries to do when he senses that something is wrong and inserts a folded postcard beneath one side of the clock-case. You can do the same, and then try bending the crutch towards the side in which you inserted the temporary wedge. Mention should be made here of the special adjustments to four hundred day clocks—though no different principle is in fact involved. These clocks employ a heavy rotating bob suspended from a fine spring—there is no pendulum rod as such—and a type of dead beat escapement is usual, as has been noted. An upright pin is attached to the pallet arbor and this is moved by a small brass fork, equivalent to a crutch. The other end of the fork is clamped to the suspension spring. The bob may turn for three-quarters of a revolution, but the fork, at the top of the turning spring, covers only a fraction of the distance, moving the pallet pin from side to side, thus transmitting impulse and working the escapement. The clock is set in beat by turning the suspension (which is a block screwed or riveted to the back plate), and very small movements are needed. The great practical difficulty lies in the flexibility of the spring, of which the length (and so also the thickness, given the need to keep time) is largely governed by the size of the clock case or covering dome. If the fork is placed too far up the spring, it will move surely but convey insufficient impulse. If the fork is placed too far down, it will flutter to and fro and the escapement will take over from the pendulum, tripping at great speed. A compromise position, a matter of millimetres either way, has to be found, and it can take a great deal of time and patience.

The beat of a balance wheel escapement is not noticeably affected by the position of the clock, but it has to be correctly set in the first place. This involves unpinning the balance spring and turning the wheel to the position which, if it were repinned, would be a central place of rest—it would move the same distance either way to lock or unlock the scapewheel. (With the cylinder escapement this normally means that the scapewheel arbor, banking pin on the wheel, and banking pin on the cock are all in a straight line.) As a rule, however, the balance cannot simply be re-pinned, for in turning it you will have passed more or less balance spring through the pinning block and so the clock will no longer keep time. Consequently, the balance must be removed and the spring turned on the staff by the same amount as the wheel was turned to set it in beat, but in the reverse direction. The turning is carried out much as removal of the spring (*see* Chapter 6), and several attempts may be needed before the balance is in beat and the balance spring pinned again so as to act freely.

## *Chapter Ten*

# BRANCHING OUT

Suppose that you are now reasonably equipped with hand tools and have had a certain amount of experience of repairs, cleaning (which should, generally, follow the repairs to a movement lest evidence of a former state is lost in the cleaning process) and assembly. You wonder about extending activities and investing in some more elaborate tooling. What should come next?

The main limitations met will probably have been in the area of the gear trains—especially broken pivots and damaged or missing wheels and pinions. To help you with these, you will need first and foremost a lathe (which can be adapted to cut wheels and pinions), perhaps a depth tool and possibly a separate wheel-cutting machine (always known as an 'engine') which may or may not cover you for pinions also. We shall look at these more expensive tools in this order because it is a rational sequence in which to buy them. A word of warning, though: I would not recommend buying a second-hand lathe or depth tool, for the risk of their being out of true is great, and it is for their truth that these tools are in fact wanted.

### *The Lathe*

Size obviously needs to be considered with some care. It is, of course, axiomatic here, as with many tools, that whatever one buys will always be just too large for one critical job and just too small for another. It is wise to think fairly big (if you have room)—that is, up to about 3½″ (90 mm). This will allow you to turn articles up to 7″ (175 mm) in diameter, i.e. with a 'centre height' of 3½″. The length of the bed is not usually of much importance in clock work, except that it is helpful to be able to have tailstock and sometimes also carriage well away from the headstock chuck. It is also worth asking yourself if you are likely to be turning weight shells or chime barrels, for a longish bed is needed for these operations, just as a considerable centre height is needed for making dials. However, generally the standard length of bed provided for a given centre height will meet all your needs.

It is wise in principle to buy apparently on the large side for two reasons. First, a bigger lathe is (or should be) no less accurate on fine work than a small one, whereas (although the scope for improvised set-ups is great in any size) a small lathe has its in-built limitations. Second, solid rigidity is vital to true work and, remarkable as some modern 'mini' lathes are, they cannot have such reliability or durability in this regard. Moreover, of course, they are frequently required to over-extend themselves, which inflicts extra strain on a tool less able to stand up to it. A long-case greatwheel is about 3¼″ (85 mm) in diameter and this may well be a size one would want to be able to accommodate, meaning a centre height of close on 2″ (50 mm) (Plate 17).

Plate 17. The Hobbymat 2½″ Lathe

Two other measurements of practical importance are less often commented upon and may even not appear in the lathe's manual. The first is the spindle (or 'mandrel') bore or capacity – that is, the internal diameter of the central tube, mounted on bearings in the headstock, on which all else depends. It might appear that for clock, let alone watch, work, only a very small bore would suffice. Greatwheel arbors seldom exceed 3/8″ (1 cm) in thickness; in any case, they are exceptional and work on them can be done by hand. But, despite this, in my experience a spindle bore of 1/4″ (6 mm) is rather a handicap, and many small bench lathes offer spindles that are smaller still. The drawback to a small spindle is that it is radical – there is no easy way round it. While most arbors can indeed be accommodated and a great deal of larger work does not protrude down the spindle at all, many pieces on which one wishes to work have a projection or piece fixed to them which ought to go down the spindle so that the other end can be worked on. For example, you may well wish to reduce a longish length of thick stock; it can be chucked, steadied and centred and may well have to be, but the occasion does arise when the ability to lose it down the mandrel would be very helpful. Again, many arbors have solid pinions on them, and it is possible to fit the end in the headstock and support the other end, for example to work on a wheel *in situ*, but you may well do a better job, and will certainly find it easier, if the pinion goes into the spindle so that you can chuck near the point of work. A verge crownwheel is a clear example of an arbor with a pinion at one end, little to hold at the other, and the desirability, for the necessary absolute truth, of truing the wheel while it is mounted on its arbor. With most lathe work alternatives and improvisations are possible, and that is true here also; but consider carefully before committing yourself to a lathe with less than 1 cm spindle bore.

Secondly, there is the width of the carriage, and also how close to each other you can get tailstock and headstock on the bed. In all lathe work the smaller the 'overhangs' – extensions of bearing or cutting pieces from their supports on the bed – the better. Overhang introduces flexing, wobbling and cross-strains. You are vitally concerned with concentricity in clock work and if, for example, a drill in the tailstock has excessive overhang, its ability to penetrate an arbor centrally is impaired. In some small lathes – understandably, often those with a large range of accessories to be mounted on them – the carriage made so wide that it is impossible to bring the tailstock drill chuck reasonably close to a workpiece in the headstock. As a result, for much of the time when drilling, the tailstock screw is advanced almost to maximum and the set-up is less rigid than it should be. At worst, the finest and shortest drills must be mounted in an additional chuck if they are to reach the headstock area at all. At this point, I should say that whether you have a hand-wheel or lever tailstock advance mechanism is relatively unimportant; the lever system gives excellent 'feel' and is quick to use, but it does not stay adjusted so well and is not so good for bigger work.

You do not need a large range of speeds in your lathe, though you will probably find it more convenient. Nor is the ease with which they can be changed a major factor. In practice you will settle down to use a couple of speeds—probably about 500 r.p.m. and 900 r.p.m.—for most of the time. A much higher speed—at least 2,000 r.p.m.—is needed if you intend to drive a wheel-cutter from the headstock (and also for grinding, but this is better done on a separate machine because the dust will damage the lathe). A much lower speed, of the order of 300 r.p.m., is needed for driving a pinion-cutter.

Many lathes are available with, and some come with, a large range of accessories, of varied usefulness for clock work. You must have at least a three-jaw chuck, a drill chuck for the tailstock, a tool holder or post, a centre, and a few turning tools. The most expensive accessory for some lathes is a vertical milling attachment. This is a vertical column, with its own powered head and chuck, to bolt onto the lathe bed for working on items attached to the carriage. The unit is extremely useful on some occasions and gets in the way on others—part of its usefulness depends on its being left in place. As a vertical drill it tends to have severe 'throat' limitations, i.e. larger work will foul the column. If you can afford it and have the space, a vertical bench drill may well be a better investment, especially if you fit it with a cross-vice so that it can be used for milling. However, for small shaping work and some drilling purposes the attachment is very convenient. The deciding factor may well be whether you consider the purchase of a separate wheel-cutting engine possible or justifiable; the use of the attachment with a cutter, at high speed and with an indexing arrangement on the back end of the mandrel, is a very reliable and attractive way of cutting wheels.

Collets and a collet attachment (Plate 18) are not likely to be included and they are an expensive addition. On the other hand, they are the most reliable way of securing concentricity when an article is fastened, then removed for some other work, then set up in the lathe again. They also have the great advantage of being considerate of what they hold; although it

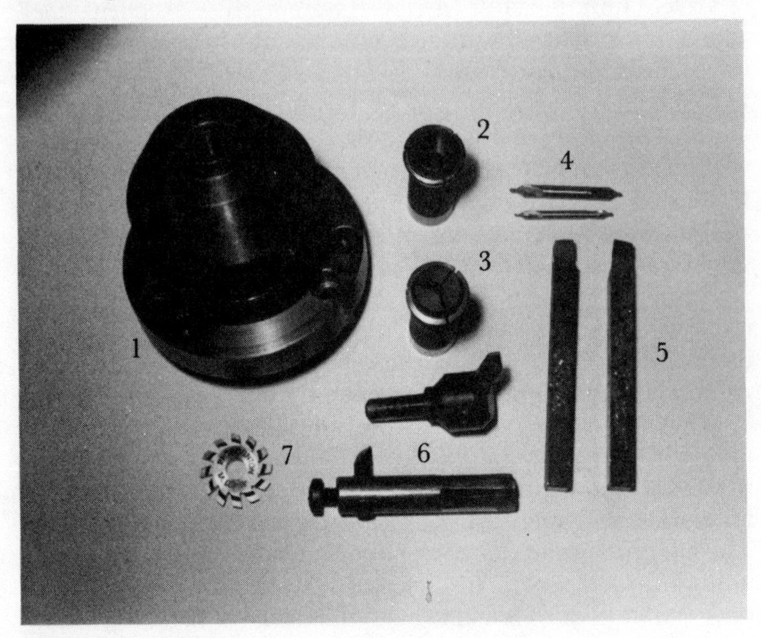


Plate 18.
Lathe Tools. 1. Collet attachments. 2, 3. Collets. 4. Centre drills.
5. Turning tools. 6. Flycutters in holders. 7. Wheel cutter

can be done, it is always a dangerous procedure to grasp pinions and wheels (by their teeth) in a chuck with pointed jaws, where any pressure can distort and break the teeth, whereas a collet will grip a pinion all round (provided it is true) quite safely. The attachments tend to be massive and the collets well made, and a considerable investment is involved when some fifteen collets, each maybe 1 mm apart, are needed; too large a collet either will not hold or will simply snap. So this is something of a luxury, to be considered as you go. Sometimes collets can be obtained second-hand and fitted with a suitable female attachment and drawbar to tighten them from behind.

*Home-made Lathe Accessories*

There are one or two accessories that it is well worth making yourself at an early stage. First, and also simplest and cheapest, is the foot-switch (Fig. 44). If you have worked without one and then acquired or made one, you will know how much difference it makes. (A clutch fitted to the lathe is the ideal alternative.)

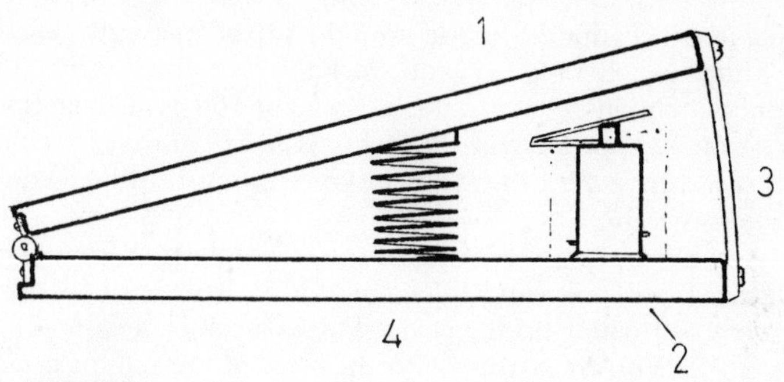


1. Pedal
2. Micro-switch and insulating tube
3. Limiting strap
4. Spring

Fig. 44. Lathe foot-switch

The on/off switch on the lathe can never be in the right place for every job and usually needs considerable finger pressure to operate it. Clock lathe work is very intermittent, although there are longer periods of work when making machine parts, and some tools (notably drills) are very fragile. Sometimes it is necessary to switch off at speed to avert a crisis or a snarl-up that would put the lathe's truth at risk. I have the same foot-switch wired into a bench drill, where similar considerations apply. You need a sprung micro-switch for mains voltage and 3 amps (at least), and of the type that is 'on' only when held depressed. Ideally it should break both power lines, but this is not essential and in any case the earth wire should not be broken. Mount the switch on the front of a base-board of about 6″ × 4″ (150 mm × 100 mm); it is best made robust and of a size that will allow it to be located with the foot, yet not overturned. Hinge a similar thinner piece of wood at the other end to serve as the pedal with which to depress it. A stiffish spring and positive action are needed—the spring in the switch will not be sufficient, so you must add a compression spring (coiled or leaf) between pedal and base. Also bear in mind that the switch will probably lie forgotten and inaccessible as swarf, unsatisfactory pivots and other things rain down upon it. Therefore attend to the insulation. My own switch—like many—had exposed soldering tags and, apart from taping it round, I protected it with a plastic cylinder—made from the cap of an empty aerosol container. Needless to say, you work the switch with the lathe or tool switch 'on', but keep the latter turned off as an over-rider when you are not busy at the lathe. It is worth first checking with the lathe maker that your motor will not be damaged by frequent switching.

I take more than a hint for another simple tool from John Wilding. It will prove indispensable and is very useful also for a first 'try' at turning on the lathe. This is a centre height gauge (Plate 19). You must turn with the edge of the tool at centre height and, if you are cutting a wheel on the lathe with a

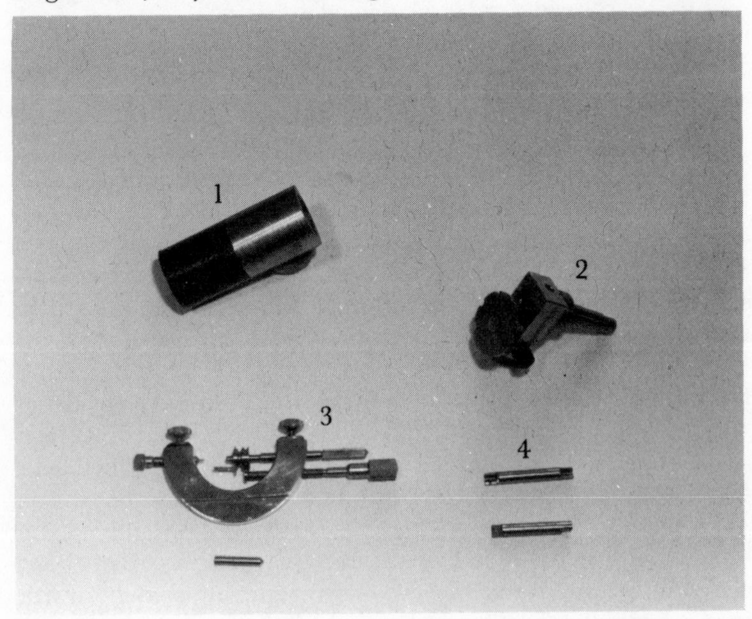


Plate 19.
Lathe and Pivoting Tools. 1. Centre height gauge. 2. Home-made jacot attachment for tailstock. 3. Drilling jig. 4. Half-runners for polishing pivots

separate cutter, you must have the cutter at centre height, or the wheel's teeth will lean. There are many other applications for a readily available indicator of height at the carriage or the headstock when it is occupied—for this purpose the centre in the tailstock may well not reach. The tool is simply a rod, thick enough to stand squarely on the bed, faced to exactly centre height. It will be easiest to make a 1″ (25 mm) brass rod, but any metal, even thick tube, will do. As a refinement, stand the finished rod on the carriage and again mark on it centre height—this is another useful measure. Turn it down, reducing its diameter by about ⅛″ (2 mm) down to this point. If your lathe is a small one these operations may well be feasible in the three-jaw chuck, but there will be considerable overhang and risk that the piece will be torn out (which could harm lathe and chuck). With a bigger centre height you will have no choice—you must support the protruding end of the rod on a centre from the tailstock, having first lightly drilled the work with a centre drill (a stubby, firm drill for this purpose which is available from any good tool shop in several sizes). Because of this support, you will not be able to face both ends completely and on this occasion it will be satisfactory to file the remaining pips flat with what has been turned.

One of the jobs that may well prompt you to invest in a lathe is pivoting. Of course, the repairer of former days used the turns quite satisfactorily and there are turns and other jigs available for doing the work by hand today. Pivots are still best polished with the arbor mounted in the lathe and with a hand-drawn bow for power. But a lathe can help, and it is virtually essential for what hitherto will have been one of the most formidable repairs—the drilling of an arbor and the fitting of a replacement pivot. Like re-facing pallets, this is a repair which has sometimes been frowned upon, but if pivot steel is used (readily available from clock material suppliers), there is no reason why it should not last. In many cases, particularly if the clock is of some age, it may well be preferable to replacing an old arbor with new metal. If, of course, the pinion is badly worn, so that again one has to choose between replacement of the part or of the whole, the preference does shift towards providing a new arbor and pinion, complete with solid pivot, and indeed towards the solid pinion rather than one drilled and fixed on. But certainly the replacement of so considerable a part in an old train is advisable only after much consideration.

Although obtaining the vital concentricity is obviously simplified by the use of a lathe, it is inadvisable simply to set up the arbor in the headstock, mark the flattened end with a fine centre drill and then drill down for the pivot (to about twice the desired visible length of pivot). The chances that the drill will flex, will not be true or will wander are far too great. Nor is it always possible to hold the arbor with the extreme end in a chuck or collet. What is needed is something that holds the blunt end of the arbor and at the same time locates the drill in its centre. In the commercial jigs graduated cones, with holes for the drills, are used and this arrangement can be made up for the lathe as well (Plate 19). Various set-ups can be devised—they rather depend on what holes you have available, or are prepared to make, on the cross-slide for holding the device in position at centre height between the two chucks. Essentially you need a plate some 1/8″ to 1/4″ (2 to 5 mm) thick, preferably of steel, but brass will do for a time, with a foot or right-angled bracket by means of which it can be mounted on the cross-slide. In this plate you drill a series of graduated holes for pivot sizes, and into their fronts, with centre drills and/or countersink bits, you put cone-shaped depressions to about two-thirds of the thickness of the metal; these receive and hold centred the ends of the arbors to be drilled. The commercial manual jig for holding in a vice is really a simple set of turns in which the arbors run on female centres at one end and on this type of cone, with the drill protruding through, at the other; the arbor has a ferrule clamped to it and is revolved back and forth by a line stretched on a wire or bone bow. As far as I know, they are made only in a size for watch and very fine clock work.

Such a power-system is advocated for polishing at least the smaller clock pivots, because there is considerable danger in mounting these pieces, with one end less than fully supported, in a lathe. Traditionally, the end to be polished rests in the trough of a Jacot tool. This is a disc, round which have been turned at centre height graded sizes of hole, the tops then being turned off, so that a series of hollows on which pivots can be rested for support remains (Plate 19). It is quite practicable for you to make such a tool (clear directions are given in John

Wilding's *How to Repair Antique Clocks, I*, Chapter 2, for one in the tailstock, and in the *Horological Journal*, June 1980, p. 16, for one on the cross-slide), but if you are new to lathe work the making of separate runners, sized as the job requires, may be more immediately rewarding. These are made of silver steel and inserted individually in the tailstock chuck. To obtain a range of sizes you will need as many thicknesses of rod (up to about 1⁄8″ or 3 mm) as you can obtain; you can then reduce some for in-between sizes. Cut the rods into lengths of about 2″ (50 mm) and then drill them in the lathe some 3⁄8″ (1 cm) deep, with drills for the thicknesses of pivot to be covered. Each runner has then to be filed to exactly half its diameter for the depth of the hole, so that half the pivot is supported in the runner, while the upper half is exposed for polishing with the pivot file. The latter is a tool which you should certainly obtain now if you do not have one already, for it is essential to pivot work. At one end it consists of a very fine file with 'safe' (i.e. flat) edges and at the other a polished steel burnisher which is rubbed across with emery paper before use to 'make' it.

### *The Depth Tool*

Although really coming into its own when a clock is being made, the depth tool is extremely useful for repairs. It is essentially an arrangement of variably spaced bearings by means of which the best depth of engagement of a wheel and pinion can be found, and by which holes exactly corresponding to this depth can be marked on plates. Its enormous advantage is that it replaces guess-work with the certainty not of general theory, but of practice in the best working of one particular wheel and one particular pinion. The calculations by circles of engagement ('pitch circles') have in practice to be supplemented or modified for a particular instance, since individual wheel-cutting varies and, where repairs are concerned, the old makers did not use exactly our sizes or profiles of wheel teeth. In any case, nothing can come closer than a depth tool to showing what happens when two gears mesh, which is commonly invisible in the depths of a movement.

There are two forms of this tool, and in addition it is possible to make up a bracket and two centres which is fixed to the lathe carriage and can carry one wheel, while another is held between the centres of the lathe. This latter arrangement does work, but it takes considerable time to make and is limited in practice by the relatively small distance between centre height and carriage, so we shall not consider it further. The classic depth tool (Plate 20) consists of two substantial brass castings, hinged and with a thumbscrew for adjusting their distance apart at the open end. At these ends the castings carry adjustable runners, with the option of male or female centres, and usually with one very large male centre. Normally the female centres support the

Plate 20. Typical Depth Tool. 1. Open runner to clear pallet crutch

parts to be tested and, when this has been done, the tool is stood on its end so that the male points, exactly the tested distance apart, can be used to mark the position for holes. If measurement from an existing hole is required, the large cone is placed in it. A very important use of the tool is for setting the best depth for an escapement, and a 'lantern' open runner is usually included, through which the crutch (passing through the line of the scapewheel arbor) can pass.

The second main type of tool is much more suitable for home construction and has several times been described in clock and model engineering journals and elsewhere (for example, *Engineering in Miniature*, July 1980; *Clocks*, June 1982; Wilding, *An 8 Day Weight Driven Wall Clock*, p. 32—these versions differ somewhat). It works on a sliding, rather than a hinged opening, principle and the major difference between it and the classic tool is that only the wheel and pinion, without their arbors, are supported and tested. The tool consists bascially of a slotted strip, at one end of which is clamped a runner with a pointed end and a boss to hold wheel or pinion (Fig. 45). To or from it along the slot can be moved another similar runner to hold the other part. This runner can be fixed in position by a thumbscrew when the optimum setting is reached so that, as before, the runner points can be used to mark the spacing on the plates.

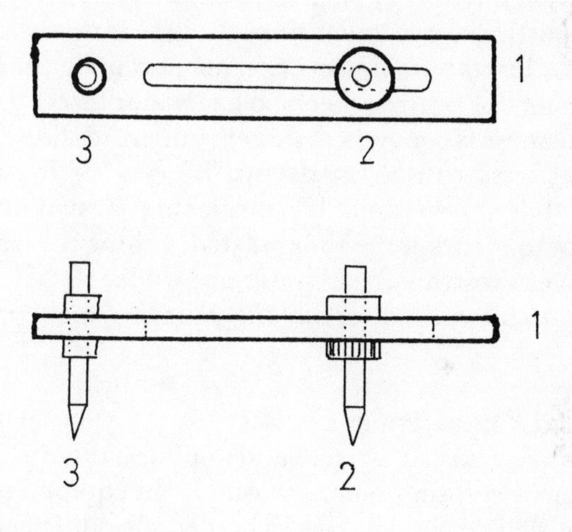


1. Slotted brass strip
2. Adjustable runner
3. Fixed runner

Fig. 45. Home-made type of depth tool (for wheels without arbors)

This second type is a sort of economy model. It is easy to make and particularly useful for clock-making, where it is convenient to test and plant wheels without their arbors. But quite a range of runners is needed and the tool is really less good for repair work where, conversely, one is often dealing with wheels that need not be removed from their arbors.

The depth tool has many uses in testing and correcting gear engagement, but one of the commonest is undoubtedly in bushing severely worn holes where the exact location of the original centre is no longer apparent. The hole can be rounded with broaches and then plugged with a piece of brass rod turned to size (rather than a bush) and riveted well into the plate. The best setting for the relevant gears is then found and the plug scribed (the depth tool being held carefully vertical) with an arc across a radial mark, so that the true centre of the hole is clearly indicated and can be drilled out. (It is a good idea to check this, trying the distanced runners in the top or bottom plate, whose holes are less worn, if at all.)

Both tools can be used for depthing escapements of the anchor type, but here again the traditional hinged type is better with an existing escapement since the arbors can be retained. It often has a limitation, however, in not accepting the longer and larger arbors of centre wheels, barrels and fusees – though, as far as thickness is concerned, larger runners to slide closely over the usual ones can be made up. In view of the amount of damage that can be done by misplacing 'drawn' and bushed holes, not to mention the time wasted, I think the classic depth tool is a very worthwhile investment, while the 'slide' type can be made up if required for making your own movements.

*Wheel- and Pinion-cutting Engines*

If you wish to cut wheels and pinions, there is no getting away from a certain amount of outlay on equipment, and you can only decide for yourself whether the independence and interest gained will make the expense of wheel-cutting facilities worthwhile.

I describe it in this roundabout way because there is no single or best way of cutting wheels and pinions. There are instead a great many set-ups used to cover certain basic requirements. Some of these are better than others for particular jobs, some people prefer one and some another, but they all need apparatus that you will probably not want, or be able, to make. As a lathe can be used for wheel-cutting and a great many other things as well, it takes priority over a separate wheel-cutter in a list of desirable tools. Similarly, as a wheel-cutting engine can usually be adapted to cut pinions, it comes before a separate pinion-cutter.

The basic requirements are easily imagined—a rotating cutter to remove the gaps and so form the teeth from what is left, and a means of turning the blank disc round step by step in degrees of a circle according to how many teeth there are. You must be able to move the cutter into the blank—or the blank into the cutter—and so you proceed round it gap by gap, and hope against hope that nothing has slipped. (If it has, your last tooth will be very fat or very thin and you will have to start again.) The size of the teeth will be small or large according to the size of cutter; too small a cutter for the blank will produce very wide teeth.

Think first of set-ups on the lathe. The blank does not need to be revolved under power, so the mandrel and chuck can carry the cutter. As cutter and blank have to be moved into each other, the blank is mounted on some device (often a vertical slide) attached to the carriage; you then have a fixed cutter position and a movable blank (Fig. 46a). However, the blank must be mounted across the plane of the cutter (which we will assume to be a sharp toothed wheel, like a commercial cutter) and on an arbor so that it can be turned for each tooth. Therefore a 'dividing attachment' is needed. This normally grips the blank's arbor at one end while its other end is fitted with either a plate with concentric holes (a 'dividing plate') representing teeth numbers (60; 72; 110 and so on), or a wheel selected for its number of teeth (an 'index wheel'). The wheel or plate can be held at each tooth or hole by a sprung lever or detent, so that the blank can in effect be advanced tooth by tooth, a cut being taken at each rest.

An alternative set-up in the lathe has the blank mounted in a fixed position on its arbor in the headstock, while the cutter is mounted, either by a small 'cutter frame', or by a bracket with bearings, vertically on the carriage—where it is driven by a belt from a separate motor—or in the chuck of a vertical milling attachment for the lathe (which has the advantage of providing an integral motor) (Fig. 46b). There are two methods of dividing in this set-up and which you use depends partly on the type of lathe. In most bench lathes it is possible to mount a dividing plate on the rear end of the hollow mandrel by means

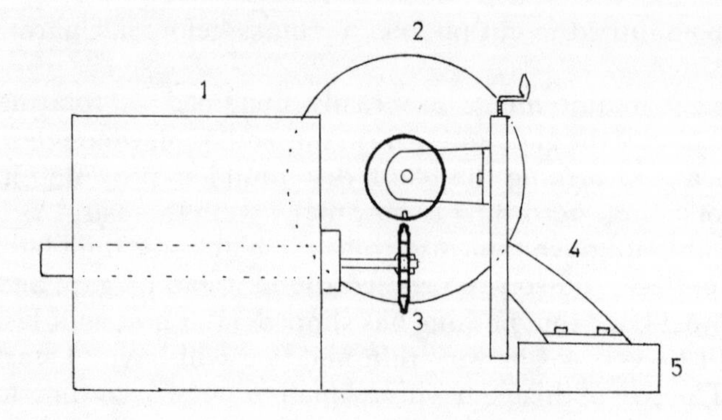


A. Cutter in headstock

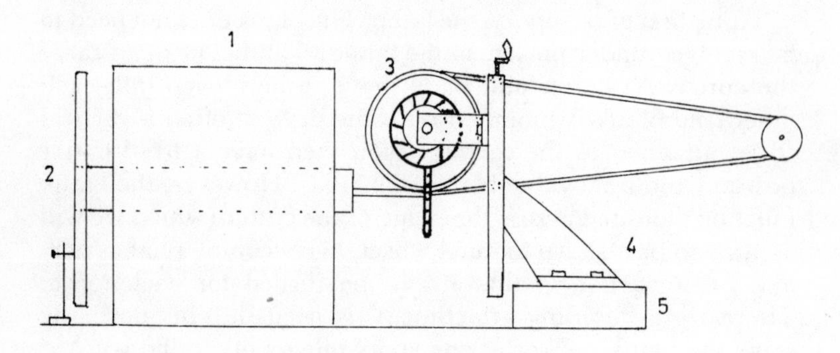


B. Wheel in headstock

1. Lathe headstock
2. Dividing plate (on same arbor as wheel)
3. Cutter
4. Vertical slide
5. Carriage

Fig. 46. Cutting gears in the lathe

of an expanding tapered plug; this is a slit tube which tightens its grip on the inside wall of the mandrel when the securing screw is tightened (Fig. 47). The attachments can be bought and are also quite easily made in the lathe; they must, of course, be made specifically for your make of lathe and dividing plate. Alternatively, if your lathe is so equipped, it is possible to use the change-wheels in the leadscrew drive, indexing the teeth in them with a sprung detent; again the detents can be bought for some lathes or can be made up. Special attachments with worm gears are also available to increase the range of wheels that can be used in this way.

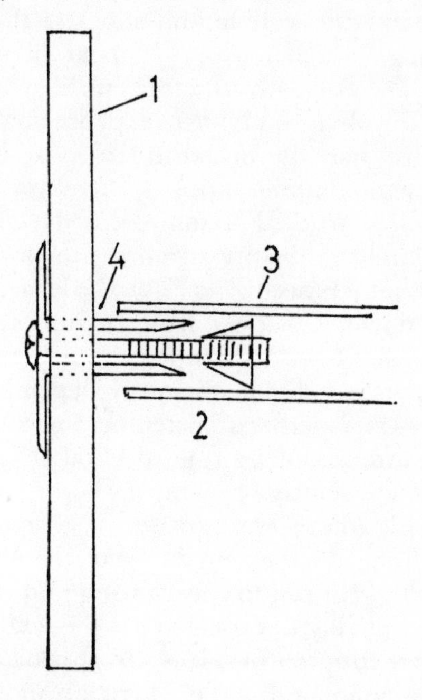


1. Dividing plate
2. Back end of mandrel
3. Threaded cone nut
4. Hollow arbor, internal taper

Fig. 47. Mounting a dividing plate on the headstock

There are pros and cons for either method, as indeed there are for using the lathe for this purpose at all. It may seem a good idea to use the lathe's motor to drive the cutter, but many lathes do not have the speed to drive a cutter properly (at least 2,000 r.p.m. for a commercial cutter, more for a home-made 'fly-cutter'). You may achieve your object after a fashion at low speeds, but the finish and shape of the teeth may be disappointing. The use of an integral pillar unit (which usually does have such a speed) to drive the cutter is highly convenient and will generally produce excellent wheels; but the fact remains that it cannot have the rigidity that is really needed and this will show if you try to cut pinions this way. Again, the taper plug fastening of division plate to mandrel should be satisfactory, but the device can wobble and slip and the slightest slip in wheel-cutting (which may not appear till the last tooth) is disastrous. The great advantage of cutting wheels in the lathe—apart from the obvious saving of expense on some equipment—s that a wheel may be turned to size as a blank and then cut as a wheel without unmounting it, thereby avoiding a common cause of lack of truth. For some, this does not overcome the irritations latent in the arrangement: the job cannot be interrupted for another piece of work on the lathe, and if some error is discovered which means starting again with a new blank, the whole set-up has to be dismantled and then put together again. How inconvenient this is obviously depends on personal circumstances—and perhaps on temperament. Finally, a 'mini' lathe may simply not be rugged enough for assured wheel-cutting, let alone cutting pinions.

Early clock wheels were made by sawing out the gaps by hand and filing up the remaining teeth. A machine was invented in the 17th century for sawing and dividing, but form tools were probably not used to cut the whole tooth (i.e. gap) until the later 18th century. It is still possible to make wheels in the primitive manner, but it is a long job and the finish is apt to be uncouth. Pinions are another matter—sawing and filing them is entirely sensible for one or two. From the later 19th century wheels could also be stamped from sheet, but pinions were extensively made from drawn steel rod—'pinion wire', no longer generally available—which was finished with files. Nowadays, however, even for small-scale production there are available separate wheel-cutting engines and pinion mills or attachments, and lengths of 'pinion blank' have come onto the market.

Wheel-cutting engines are generally of a well-proved pattern. The blank is mounted horizontally on a vertical arbor to which the dividing plate is fixed below (Plate 21). From the bed in which this arbor rides ascends a vertical column on which is mounted, on a slide, a cutter frame with a lever for moving it up and down, with a pulley and simple drive belt to a detached motor. The column itself is on a slide so that the cutter can be

Plate 21. A Wheel-cutting Engine.

moved into the work. The speed will be 3,000–4,000 r.p.m., though it can be altered by changing either the motor or cutter pulley. All is simple and very firm. There may be the disadvantage of mounting the blank on an arbor different from that on which it was turned to size; but even this is overcome in some machines which, for example, arrange for a standard transferable Myford arbor to be used. However, a change need not be detrimental, and in any case it is much easier to have another go here than it is on a lathe set-up.

Pinions can be cut by these machines with attachments, as they can on the lathe. However, it is laborious and not entirely satisfactory. This is because large amounts of a harder metal have to be removed from what becomes quite a frail article. It is worth trying, noting the need for a speed of only about 200–300 r.p.m., and devising stout steadies for the unsupported end of the pinion, but it is generally better to buy blanks, have pinions made, use lantern pinions in your own designs (these are not difficult to make) or, if use will seem to justify the expense, buy a pinion mill. The Chronos mill (Plate 22) is ruggedly built and

Plate 22. The Chronos Pinion Mill

versatile and, after the other methods, one uses it with relief. It holds one end of the pinion arbor in a collet or chuck attached to a simple dividing plate while the other end is mounted on an adjustable centre along the tool's bed. This bed is fitted with slides so that it may be moved up and down (in relation to the cutter which is above) and travel sideways; the cutter must, of course, progress *along* a pinion, rather than straight through as with most wheels, and indeed several pinions of the same size can be produced by making a long one and cutting it up.

*Wheel- and Pinion-cutters*

No matter what set-up or machine you adopt, it will, of course, require a cutting head. Commercial cutters in high-speed steel are excellent but, alas, expensive, partly because of the range that a repairman may need. Theoretically, teeth of a set of 60 are of a different profile from teeth of a set of 48 in the same size, and so on. In practice a single cutter can be used for any tooth number that you are likely to meet in a size, but the number of sizes still means a big outlay if you are to cover most contingencies. For pinions, angles vary more, and more cutters are needed.

Modern multi-tooth cutters are sized on the module basis. The module is the pitch diameter per tooth of the gear—that is, diameter at point of engagement (i.e. less than the full external diameter) divided by the number of teeth, with the answer in millimetres. Clearly it is of no use until multiplied by a specific number of teeth, when the product will be the pitch diameter of a circle needed to accommodate so many teeth of that size or module. To give you an idea of the range, it may be said that carriage clock wheels tend to go down to about 0.35 module, bracket clocks may be between 0.5 and 0.7, whereas lantern and long-case clocks may be between 0.75 and 1.0. Plate 23 reproduces full size a simple gauge showing tooth sizes (and numbers for this diameter) and modules in 0.05 mm steps from 0.35 to 1.0.

Plate 23. A Wheel Tooth Gauge

Even if we have ample finance, commercial cutters are not necessarily the answer to repair problems. The old craftsmen did not use mass-produced cutters sized in this way and, although a wheel produced by them may be satisfactory in an old train, it may well be better as well as cheaper to make up 'fly-cutters' (bearing in mind that they need to be driven at at least 6,000 r.p.m.). You can make them direct on the grindstone or you can make another cutter from which to shape them on the lathe—the latter method produces cutters giving a better finish. Cutters can be made from silver steel, from old files softened and then re-hardened, or from the tool steel conveniently made in ¼″ square section by Eclipse. Silver steel is best for pinion-cutters.

In profile, a cutter exactly matches the gap between two wheel teeth. The difficulty with grinding and filing cutters to a pattern, or making them to size, is in getting both sides the same and so avoiding lop-sided and irregular teeth. There are two ways of avoiding this in the lathe. The first method is to grind or file a cutter to use as a form tool in turning the end of a steel rod (to be your wheel-cutter) of the desired maximum diameter. This form tool is thus the profile of one side of a tooth but, since you turn the rod, the profile appears in the round on the wheel-cutter (Fig. 48a). Take great care with the curve of the cutter, slope well back away from the edge for relief, and polish it with an Arkansas stone when finished. A cutter can also be made from part of a file that is softened at bright red heat, shaped, and then hardened by being plunged red hot into cold salt water. It is then cleaned and polished and tempered to a light straw colour. When you have used this tool on the embryo wheel-cutter in the lathe, the latter will be cut to the shape of the space between wheel teeth all round. It is then filed down flat to exactly half its diameter and relieved at the front end, the cutting face.

The other method of obtaining complete symmetry is to use sections cut from a steel blank whose thickness is the maximum width of the tooth gap (Fig. 48b). The blank is again mounted in the lathe, on a stout arbor, and turned with a similar form tool of half profile, this time from each side. You then have a steel disc milled at the edge to tooth-shape. Cut the shaped blank across its diameter and then cut or file away the outside of the circle; this gives you two cutters, and their edges, with the blank's original curve, already have the necessary relief or

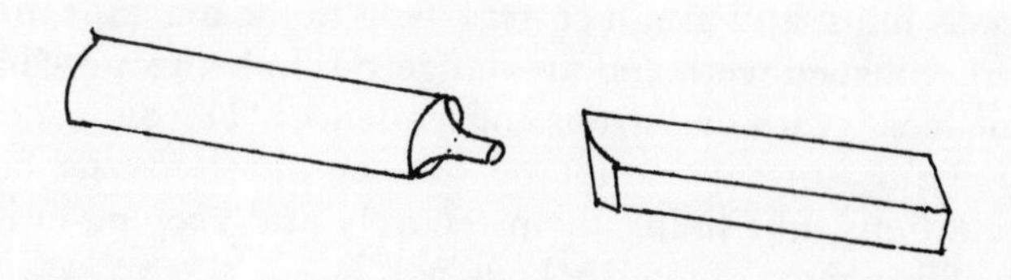

A. Turned rod type

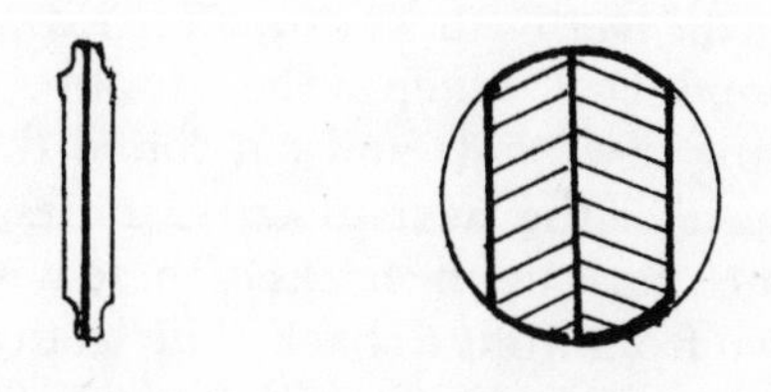

B. Cutters formed from disc

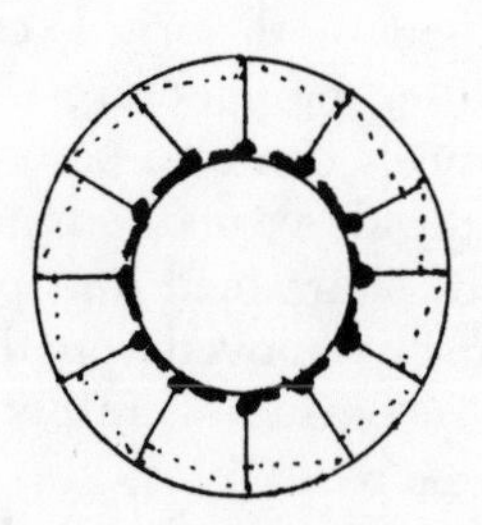

C. Multi-tooth type

Fig. 48. Home-made wheel-cutters

clearance. They must, of course, now be hardened and polished. The polishing of fly-cutters is of the utmost importance as blunt and rough cutters bang in the machine, produce poorly finished teeth and are dangerous in being more likely to come loose, with disastrous consequences. The danger of literally flying cutters cannot be over-emphasised and sensible precautions, like fixing them securely and keeping out of the line of fire, must obviously be taken.

There are also ways of making your own multi-tooth cutters, and I would particularly commend that suggested by Chronos Ltd (Fig. 48c). Make a milled steel blank to tooth profile (as above) and mount it on a 12 hole dividing plate (or one with a multiple of 12, such as 72 holes). With a rotary slitting saw cut 12 radial lines for the teeth – these can, more slowly, be marked out and cut by hand. Drill a hole at the bottom of each tooth to accept the piercing saw blade and cut round the inner circle almost to the start of the next tooth 12 times, then with a hammer and punch close up the backs of these slots so that each tooth slopes down from front to back. The teeth are then filed and polished where exposed to form the cutting faces. Incidentally, when hardening cutters it is a good idea to wrap them in soft iron wire and to smother them in soap (the soft soap off the back of a used block of hand soap will do) – the iron helps preserve their shape while the soap makes them easier to clean and polish after hardening. Such cutters can be sharpened, as can the commercial ones, since their tooth form is constant. Suitable stones can be had from cutter manufacturers.

Cutters for ratchet and scapewheels can also be bought or made. If you make them by any of the above methods, take care that they are thoroughly relieved, sharpened and polished. Even so, a good deal of metal has to be removed from the wheel blanks and it is best to go round several times rather than to hope to slice out a whole tooth in one pass.

### *Cutting Wheels and Pinions*

Some information on sizing will be given in the next section. The procedure for cutting has already been indicated and needs little elaboration. Probably the most crucial step is in centring the cutter. This is particularly difficult, but absolutely vital, with pinions, where the diameter is so small. If the cutter is not properly centred, leaning teeth will result. The cutter may be deliberately offset to produce raked ratchet teeth but otherwise, once it is centred and if all your cutters are the same thickness, the position need not be varied. In some cases it will be possible to scribe a line of centres from the blank's arbor to the cutter, or to mark how far the cutter arbor protrudes from its bearing, and this is a useful guide. Try out the setting on an old wheel, blank or rod rather than mutilating the piece to be worked. Be particularly careful if the blank is in the headstock and you use a vertical slide or milling device to house the cutter (Fig. 46b); you cannot use the tailstock centre as a guide, and this is an occasion when your home-made centre height gauge will be handy. Observe the angles minutely or try out several teeth on scrap; it can happen that although a single tooth looks straight, a batch reveals a significant error in centring.

You then have to establish the desired depth of cut. This is done by gradually advancing the cutter in two gaps round one tooth until it is perfectly formed. It is best to leave the very slightest flat—at any event, do not risk cutting too deep or your teeth will be too small. Once you have found the required depth you *must* lock the cutter arbor in position or you will find that you are getting inconsistent teeth. Most clock wheels can be cut with one pass of a sharp tool, but greatwheels and ratchets may require several cuts. Pinions certainly will. They are difficult, but not impossible, to cut with fly-cutters. The Chronos pinion mill has already been mentioned, but here you need commercial cutters and this means two or three to each module size to cover 6, 7, 8, 10 or 12 leaved pinions. A slow speed is needed and either the free end must be supported, as in the mill, or some sort of support beneath the line of cutting has to be devised—a solid wedge is quite satisfactory. Pinions being cut become very hot; both to cool them and to lubricate the cutting, obtain a good supply of soluble cutting oil, which can be kept handy in an old 'squeezee' bottle.

Contrate wheels and verge crownwheels appear complicated, but there is no reason why you should not produce them with comparatively meagre equipment. The blanks are cut from brass rod somewhere near the right diameter and turned down to exact size; this size is often not very critical, as is the case with ratchet wheels as well, but it can be worked out by multiplying the base width of the teeth by the number of teeth required. When it has been turned to size, the blank is drilled for the wheel-cutting arbor (or its eventual arbor or collet if these can be used in the engine), following which it is parted off from the stock (which may have been held with a tailstock centre although this can be left until later). You can mount the blank in the 'wide' chuck for boring, but it is better to make a close-fitting recess in a wooden blank and to chuck that, with the wheel blank pressed or stuck in. Provided you have already drilled, the boring can be carried out with an ordinary turning knife tool. It is easy to over-estimate the required thickness of both walls and bottom, so watch them carefully, bearing in mind that the bottom will probably be crossed out if it is not to be very heavy. One of the problems of contrate wheels in operation is that they do not present a distinct line to their pinions. The thicker the wall, the more acute this difficulty becomes. However, if the wall is too thin, the teeth will be too weak and can also bend or break when being cut, especially if the cutter is home-made and not as sharp as it ought to be. Therefore it is usual to taper the wall of a contrate wheel, making the teeth thicker nearer the base than at the top. (The depth or height of the wheel is not usually absolutely critical, since verge and contrate wheels are generally provided with pivot end screws to adjust their depths of engagement—which *are* critical, to the point of capriciousness.) Be very careful towards the end of the boring that you do not slice into the bottom near the walls—the bright brass gives out deceptive images; I usually bore leaving a thick wall and then thin it and true it up in a separate operation. Once you have bored the blank, cutting proceeds as usual except that, if your tooling will permit, it is better to come up at these wheels from below rather than down on them from above, which is more liable to knock the wall over; cutting set-ups can usually have their direction of rotation reversed by crossing over the drive belt.

At this stage it is appropriate to mention that the big recoil of a verge escapement means that its contrate wheel alternates between being driving and driven. This further complicates this type of gearing, for one can find one has cut a wheel which is acceptable in one direction but which jams in the other. The practical solution lies in making the blank a little oversize—this simply means that it rides further up the pinion and does not affect the depthing—since larger than usual gaps between teeth help considerably. When cutting the wheel, after the teeth have been cut, move it very slightly sideways and go round again, thus enlarging the spaces. You will then have to touch up the tooth tips with a file. This is necessary with verge teeth anyway, since it is unwise to cut them to a sharp point on the machine, though they should after filing still be left very slightly blunt. (As always, never touch radial fronts of teeth once cut, but only their backs.)

Contrate and crownwheels are so finicky that it is especially important that nothing should be left to chance. As they may well be slightly distorted during the cutting, it is essential to take them back to the lathe, remount them in the wooden chuck block and true the central hole with a sharp fine tool, or ream it. The horizontal truth of the verge crownwheel is a prerequisite for the escapement to work; spin the wheel and, using a scriber or tool in the rest as a gauge, observe whether all the teeth are of exactly the same length. If they are not, true the wheel by dressing with a slipstone in the tool rest when the wheel is mounted and colleted on its arbor. Crossing out (i.e. making the spokes) is not attempted until this stage, since it weakens the wheel. When it has been done, truth should be tested again.

There are two other things I should mention in connection with wheel-cutting, and these are repairs to wheels and crossing out. While it may be best to repair a wheel by filing it out and inserting a new tooth (*see* p. 140) which is then filed to correspond, cutting facilities may enable a better repair, especially if several teeth are damaged (as tends to be the case). By close comparison with the existing teeth you have to select the appropriate cutter or else make a fly-cutter; it must be remembered, of course, that no modern cutter may provide a reasonable fit. Saw out the broken teeth obliquely, as when doing a hand repair, and soft solder (do not braze) a slip of brass into the gap, ensuring that the outside edge is never below the outside circumference. Then fit up the wheel for cutting and use a dividing plate with the number of holes for the wheel teeth. Set the wheel so that the cutter is opposite the gap of the last undamaged tooth and advance the cutter into this gap until it is an exact fit—very great care is needed in doing this, so now take the wheel and dividing plate round to the other side of the gap and ensure that the cutter fits the last tooth gap there as well. You can then cut round the partial blank. This procedure does not take long and can be well worthwhile even if only a single tooth has to be replaced.

Amateurs tend to have difficulty in cutting spokes (crossing out) and this is not surprising as it is largely a matter of practice. You need a proper piercing saw, not a fretsaw (whose balance is wrong), and a really stout slotted table; the latter is not difficult to make, but an excellent commercial one from Chronos Ltd is not very expensive (Plate 24). An opportunity for gaining experience with the saw, which can be arranged more easily than crossing wheels, is designing and cutting hands out of mild steel or gauge plate of about 2 mm thick; you can make useful hands of better quality than the stamped ones often supplied, and you will find that crossing wheels is comparatively easy work when you come to it. In either exercise, err on the fine side in choice of blade; a heavy blade has to cut more metal and it will break just as easily if you twist it (the main

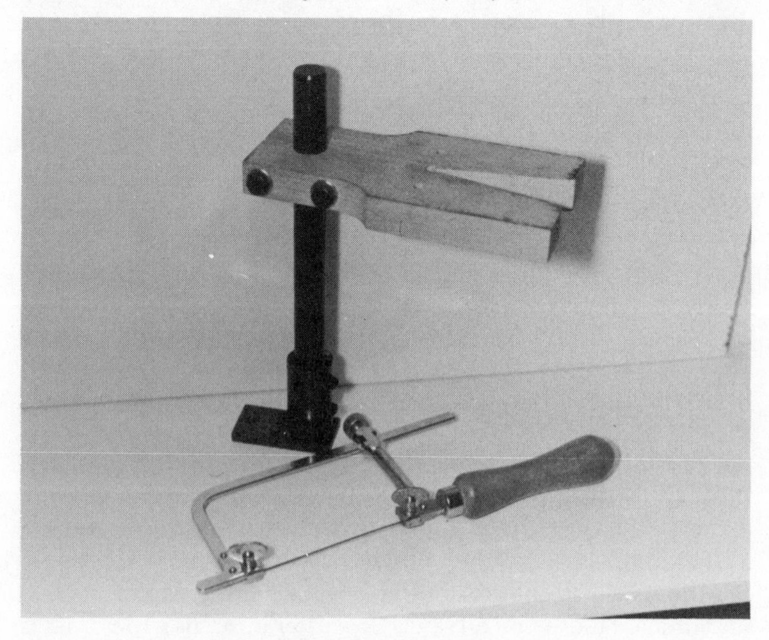

Plate 24. A Piercing Saw and Chronos Table

cause of fracture). In marking out crossings, I would strongly recommend John Wilding's gauge (*see A Weight-Driven 8 Day Wall Clock*, p. 24), but it is perfectly possible to use a scriber in the lathe tool post with a dividing plate on the mandrel, to adapt the wheel-cutting engine, or indeed to do the job virtually free-hand. Incidentally, some small wheels (including crown-wheels) are crossed into three only and these divisions can be obtained by wedging in turn each jaw of the three-jaw chuck. You may have to drill all sharp corners to give access to the saw blade, but with practice only a starting hole will be necessary. Take great care with this because a drill hole that is placed slightly crooked in metal that should be left intact is very difficult to erase by filing without damaging the proportions of the crossings. Half round or (flatter) barrette and triangular files are the most useful for finishing. Take the side off some files so that they are 'safe' from rubbing adjacent metal—they can be ground off. Remember that the saw should be vertical in use, or you will have sloping edges which will be difficult to correct by filing, and of course the wheel should look the same from both sides. The crossings and circumference can finally be burnished by rubbing with a large needle, using considerable pressure. The edge of a cutting broach is also useful here, both with new and existing wheels.

### *Sizing and Missing Wheels*

The modular basis of sizing has already been outlined, being based on the circle of engagement or 'pitch' circle. If you multiply the module size by the number of teeth required you will obtain the pitch diameter of the blank. However, what in practice you need in order to cut the wheel is the overall diameter, i.e. including the tips ('addenda') of the teeth. This is arrived at by adding an addendum factor to the formula just given, and this factor is a constant for a particular make of cutter, though it varies for pinions according to the number of leaves. The factor is normally in the region of 3.1 for wheels and 1.7 for pinions. If, therefore, we take the gauge illustrated in Plate 23 and consider 0.6 module for the diameter of 51 mm, we find by the calculation (turned round)

$$\frac{51-3.1}{0.6} = 79.8,$$

that 80 teeth of this module require a blank just over 51 mm in diameter. Normally, of course, it is in the form $(80 \times 0.6) + 3.1 = 51.1$, since you know that 80 teeth are required. Exactly the same procedure, apart from a lesser factor, is used in calculating a pinion blank. As has already been suggested, you may well find that such measurements do not work out exactly on an old clock, but in such an instance you have sufficient information—the calculated number of teeth, the evident total distance between wheel holes (comprising the sum of two pitch circles), the size of the pinion, the gauge of neighbouring wheels (one of which is likely to be the same)—to come very near to what is required, although it is possible that more than one trial will be needed. Incidentally, it is vital, in calculating for a contrate blank, *not* to include any addendum allowance; the tooth tips do not contribute to the blank's size.

As can be seen, a crucial factor in wheel calculations is the number of teeth. That is, as was said earlier (*see* p. 16), essentially a measure of size. This is so central a matter that, while the information is widely available elsewhere, a little must be said here to indicate how one may go about identifying a missing wheel. We shall consider representative situations—a wheel in the time-keeping train, a wheel in the duration train, a wheel on the striking or chiming side, and finally the gearing of the external countwheel on English clocks (since this is quite often missing).

Assume first that you have a long-case clock with a missing scapewheel. These calculations must always start from a known factor or objective—the right-hand side of the equation. If the time-keeping train is concerned, this datum is the period of the oscillator. In some cases this will have to be deduced from such of the train as you have, but for a long-case clock it will be one second (or thereabouts, in thirty hour clocks). With a bracket clock you will have the evidence of the case size and mounting if you do not have the pendulum. Pendulum sizes of French movements are often marked in decimals on the backplates—this will be in centimetres or in French inches (which were 1.066 as big as English inches). Several handbooks tabulate pendulum lengths and frequencies. You then set out the count, starting with the centre wheel:

| wheels | 60 | 56 | scapewheel |
|---|---|---|---|
| pinions | | 8 | 7 or 8 |

Clearly the second wheel revolves in 8/60 of an hour, 8 minutes. The scapewheel will revolve in 1/8 of this (1 minute) if the pinion is of 7, or in 1/7 of it (1 minute 8 seconds) if the pinion is of 8. We can be sure, especially if there is a seconds hand, that the pinion should be of 7 and that the scapewheel is to revolve in 1 minute. This requires 60 swings of the seconds pendulum, but the escapement releases only half a tooth at a time, so there will have to be 30 teeth on the scapewheel (which is normal in these clocks).

As far as duration goes, the greatwheel of such a clock is unlikely to be missing, but the same principles apply to other clocks, especially eight day spring-driven movements with an intermediate wheel. For example, you might be lacking the intermediate wheel of a French clock but have the barrel with 80 teeth and the centre wheel pinion with 8 leaves. The datum here is uncertain, being simply a reasonable number of revolutions of the barrel in a week, but, of course, in practice you will know the approximate size of the missing wheel from the plate holes and this will limit the options. The total space between holes will give you the sum of the wheel and pinion pitch circle radii and, measuring with a vernier caliper, you can deduct the circle of the pinion. If you can then establish the module (or equivalent), you can arrive at the tooth count, as indicated above. Taking the count alone, however, the existing ratio from greatwheel to centre pinion is 80:8 or 10:1, which means that the barrel would revolve once in 10 hours, or 19 revolutions would be made in a week. The intermediate wheel is provided because clearly this would be impracticable. It may revolve once in, say, 8 hours, which will give it 64 teeth ($^{64}/_{8}$), when the greatwheel will revolve 2½ times a week if the intermediate pinion has 8 leaves. Such a count is common in French clocks, which often run for much more than a week and where the intermediate wheel could have been dispensed with if a barrel of higher count had been used.

The standards are obviously different with striking trains and there are several requirements. They are that the warning wheel must always stop in the same place, as must the locking wheel, so that there must be whole ratios here. Since a blow must be made for each revolution of the locking wheel, the number of pins on the pinwheel must equal the ratio of the locking wheel pinion to the pinwheel teeth that drive it; for example, you could have 10 pins on an 80 tooth pinwheel which engages an 8 leaf locking wheel pinion, but you could not have 72 teeth on such a pinwheel, nor, with 80 teeth, could you have a pinion of 6 or 7. The teeth of warning wheel and fly are governed only by the speed of striking required and the space available. Most chime ratio wheels (outside) have a spacing function as well as transmitting power to the chime barrels. Usually they give the countwheel a ratio of 2:1 to its barrel, since the barrel normally revolves twice in an hour and the countwheel (on the front plate) only once. The greatwheel must, through any intermediate wheel, again have a suitable ratio to the pinwheel to give duration. This is often 60:1. Where there is no intermediate wheel, as in a long-case clock, a ratio of about 10:1 is usually adopted.

All this information should enable you to solve problems with missing strike and chime wheels, but the case of external countwheels, most often on thirty hour long-case clocks, is a little different. Here we are concerned with the total ratio through from locking wheel to the wheel attached to the countwheel which, in a thirty hour clock, is driven by a pinion mounted loosely on the arbor of the greatwheel (which is also the pinwheel). This ratio must equal the number of divisions on the countwheel—78 on an English one and 90 on a French one. As before, the ratio of the pinwheel to the locking wheel must be the same as the number of pins. Typically, there may be 13 pins, the locking wheel may have a pinion of 6 and the greatwheel (pinwheel) be of 78. This ratio of 13:1 leaves us 6:1 to make up and this may be done in various ways so long as the wheel attached to the countwheel has 6 times the number of teeth as on the greatwheel external pinion. These pinions are often not pinned on, but merely held in place by the overlapping countwheel. Alternatively, the pinion was simply cut out of the extended greatwheel arbor and might have as few as 4 leaves in the shape of protruding pins or a rudimentary lantern pinion.

When you have established the number of teeth for the missing wheel (or pinion, where of course the methods are the same) and can start to make a replacement, you need, as when repairing a wheel, to match the teeth to a module cutter or to make a cutter to size. This size will almost certainly be the same as that of one of the adjacent wheels. To establish the position, however, divide the missing diameter (i.e. pitch diameter) by the number of teeth and this will give you a figure in the module range, though probably not an exact module. For example, it might be 0.635, where the nearest half module size (0.65) would be quite satisfactory. This size can be checked with the gauge in Plate 23 and a cutter either bought or made. If you order a wheel-cutter, high-speed steel is now generally used (there used to be an alternative of carbon steel) and the holes are standardised at 7 mm, so you need specify only the module; but if you order a pinion cutter you must specify the number of leaves as well as the module since, as explained above, angles vary significantly according to the number of leaves.

## GLOSSARY OF TERMS

*(This Glossary excludes certain words of narrow usage which are defined as they occur in the book).*

ARBOR The horological term for an axle or spindle carrying a wheel or pinion.

ARMATURE The iron component which moves as a result of magnetic attraction, particularly the rotor of an electric motor and the moving part of an electric bell, buzzer or similar arrangement.

BALANCE SPRING The fine, flat coiled spring, popularly known as the 'hairspring', which largely determines the period of oscillation of a balance wheel of a watch or clock.

BARREL A hollow cylinder, usually of brass, turning on an arbor. Mainsprings are often contained in barrels and the line of a longcase clock is wound round one. The chiming barrel is a cylinder with pins or other projections for operating chime hammers.

BEAT The oscillations of pendulum, balance wheel, tuning fork or quartz crystal, with particular regard to their regularity when heard; thus a clock which does not tick evenly because its pendulum or balance is crooked is said to be 'out of beat'.

BEZEL The circular metal flange into which a clock glass is set.

BRACKET CLOCK A term used, particularly of English clocks of the 17th and 18th centuries, to distinguish small standing clocks from hanging wall clocks and longcase clocks. In fact, only some so called had brackets made for them or were mounted in this way. Later, less imposing, clocks are often referred to as 'mantel clocks'.

BRUSH A springy contact resting on the commutator in electric motors and conveying current to a wire-wound armature. Clock synchronous motors do not have wire-wound armatures or brushes.

BUSH A brass tube inserted in a pivot hole to correct for wear.

Clock bushes amount to substitute holes which are fitted carefully to receive the pivots and then riveted into place in the movement plates.

CANNON PINION A tube, terminating in a pinion (wheel) against the front plate, which fits over the centre arbor of many clocks and watches, being friction-tight with that arbor and carrying the minute hand.

CARRIAGE CLOCK A small clock with brass pillared case and glass sides and top, through which the platform escapement can be seen. (Less often, the sides may be enamelled or the top solid.) The origin of the term is debated, but it plainly refers to the portable nature of these clocks which were first made in the early 19th century and the majority of which are French. Many replica carriage clocks have been made in recent years, especially in England.

CHIMING The term is generally restricted to sounding at the quarters (or at other times, but not simply at each hour) on more than two gongs or bells, and with the use of a separate gear train for this purpose. 'Quarter striking' involves the use of the same train for striking hours, but 'quarter chiming' involves a separate train at the quarters.

CLICK The pawl which, with the aid of a spring, prevents the backward movement of a ratchet wheel; in particular, the pawl which prevents a mainspring from unwinding save by turning the wheels in a gear train.

CLICKWHEEL The ratchet wheel on a mainspring winding square which, with click and clickspring, prevents the mainspring from unwinding save by turning the gear train.

CLUTCH In this book used to describe the friction drive arrangements by which the hands and their associated wheels can be turned independently of the main gear train for the purpose of setting clock hands to time.

COCK A bracket—particularly the bracket by which balance wheel or pendulum clock pallets are mounted.

COLLET A collar or washer, such as that by which a wheel is mounted on its arbor.

COMMUTATOR Segments of a cylinder on the armature of an electric motor by means of which, in conjunction with the brushes, current to the armature can be switched on and off according to its position in relation to the field magnet.

COUNTWHEEL A wheel moving forward at regular intervals which determine the operation of some other part of the movement. Thus the countwheel may dictate how many blows are struck or chimed, and the countwheel of the Synchronome clocks counts the seconds in a half-minute period upon which an impulse to the pendulum is given and the dials are advanced.

CRUTCH The forked lever which connects the pallets (on whose arbor it is attached) to the pendulum rod in a pendulum clock.

DETENT A metal projection or spring which may set off or stop movement of a part.

ESCAPEMENT The device by which the powered gear train is constrained to move at a constant rate dictated by the vibrating pendulum or balance wheel. Essentially it involves the periodic obstruction and release of teeth on a gearwheel (the scapewheel). A second function of the escapement is to give the balance or pendulum the impulses necessary to maintain vibration.

FLIRT A small piece moving with a jerk and usually impelled by a spring; particularly such a piece which, released by the lifting pin, flies across a movement to release a striking or chiming mechanism.

FOUR HUNDRED DAY CLOCK This is the type of clock, also known as an Anniversary Clock, which runs for a year on one winding, and has a rotating torsion pendulum on a long suspension spring, usually vibrating some eight times in a minute.

FUSEE A tapered and grooved barrel used for equalizing the force of a mainspring at all stages of its unwinding.

GRANDFATHER CLOCK Strictly, a 'longcase clock'. The term 'grandfather clock' derives solely from the popular American song 'Grandfather's Clock', which was written in or about 1878.

GREAT WHEEL The first wheel in the going or striking train, nearest to the source of power.

HAMMER TAIL A projection from the hammer arbor, or an extension of the hammer itself, by means of which pins of the pinwheel lift the hammer for striking.

HERTZ The term (after the German physicist Heinrich Hertz)

now generally used to denominate the number of oscillations in a second (formerly 'cycles per second').

IDLER WHEEL A gearwheel whose only function is to cause a reversal of motion in a gear train, or a wheel, or pawl standing ready for occasional use in a movement.

IMPULSE The energy which must be delivered to an oscillator to keep it vibrating. It is one of the functions of the escapement to give impulse to pendulum or balance wheel. In many electric clocks impulse is given by an electromagnetically operated mechanism.

ISOCHRONISM The property of vibrating over a greater or smaller distance (amplitude or arc) in exactly equal periods of time. Properly adjusted pendulums and balance wheels approach the condition of isochronism; variations in the size of their swing have relatively little effect on their timekeeping.

LIFTING PINS Pins spaced round a cannon pinion on the central arbor, corresponding to the minute hand position at the hours or quarters. The lifting pins operate the flirt or lifting piece to initiate striking and chiming. Sometimes a shaped cam is used instead.

MOTION WHEELS The gearwheels and pinions, generally between the dial and the front plate but often between plates, by which the revolutions of the central arbor are reduced to one revolution in twelve hours for the hour hand, with all hands turning in the same direction. Motion wheels are outside the main train which runs from mainspring or weight barrel through to the escapement.

OSCILLATION A cycle or double vibration—a swing of the pendulum or balance to both sides and back to the starting point.

PALLET A projection, on a vertically moving part, which engages with a rotating part; or a rotating projection engaging with a vertically moving part. Principally escapement pallets, which engage with scapewheel teeth, and gathering pallets which gather up striking and chiming rack teeth one by one.

PAWL A pivoted pointed lever which rotates a ratchet wheel in one direction or which allows rotation in one direction only, e.g. mainspring winding.

PINION A small gearwheel, of not more than 20 teeth and most often of from 6 to 12 teeth. Pinions may be in one piece (of metal or synthetic material), or steel rods in brass endpieces ('lantern pinions'). Generally they are solid, made of steel and may be part of their arbors driven on to them. A pinion is usually driven by a larger wheel, except in the motion work and some electric movements.

PINWHEEL Also known as 'hammer wheel'; the wheel with regularly spaced projecting pins which operate the hammers of a striking or chiming clock.

PIVOT The finely turned end of an arbor which runs in a hole in the movement plate or a special cock.

RACK A vertical or horizontal series of gear teeth with which a pinion engages, moving it up and down or sideways. Most commonly, striking and chiming racks (in fact, segments from large circles) gathered up a tooth at a time by a gathering pallet rather than a pinion.

RATCHET A saw-toothed wheel restrained by a pawl or click from moving in one direction.

RATIO WHEELS The wheels, most often fastened to the outside of the backplate, by which the revolutions of the chime barrel (and so the chiming sequence) are matched to the periodic running of the chiming train at the correct times indicated by the hands.

REPEATER A clock or watch which, on pressure of a button or pull of a cord, repeats the last hour struck. More complicated mechanisms repeat quarters and hours or indicate the nearest minute. In some cases automatic striking is not present and the clock or watch will sound only on demand in this way.

ROTOR The revolving part of an electric motor, particularly the unwound armature of a synchronous motor.

SCAPEWHEEL The wheel, furthest from the source of power, on which the pallets of an escapement act. The shape of the teeth varies according to the type of the escapement but has roughly the profile of a ratchet tooth.

SHAKE The word used for freedom of a moving part in relation to a fixed part, most often of pivots in their holes.

SNAIL A cam, in shape like the profile of a snail's shell, which is used to regulate the number of blows struck at each hour

(hour snail) or quarter (quarter snail). Often the twelve steps corresponding to the hours and on which the rack falls are cut on the snail, and the quarter snail is always in four steps.

STAFF A lathe-turned arbor, most commonly that of the balance wheel or pallets.

STAKE A tool, usually of steel, with graduated holes. It is held in a vice, allowing an arbor through the appropriate hole, and facilitates many operations, particularly punching and riveting.

STATOR The static field coil or magnet of an electric motor.

STRIKING Sounding of the hour, and possibly the half hour and quarters from the same train, on bell or gong. *See* Chiming.

STUD A short metal rod of which the tip is threaded to be screwed into a plate or similar piece. Many of the parts associated with striking and chiming are mounted on studs screwed (or, more recently, driven or riveted) into the front plate.

TRAIN An arrangement of pinions and wheels which act on each other in series.

TRANSISTOR An electronic device capable of oscillating, or of acting as a relay or switch, when part of a suitable electrical circuit. Transistors can operate on a very low current, generate negligible heat, and are very small, so that they are indispensable in modern electronic circuits.

VIBRATION The swing of a pendulum or balance from one side to the other; half an oscillation.

VIENNA REGULATOR A weight-driven wall clock with a pendulum. Strictly reserved for high quality clocks of this type made in Germany and Austria in the 19th century, having no striking and being clearly directed towards refined time-keeping; in practice the term is used of good weight-driven pendulum wall clocks, usually with glazed sides and brass-cased visible weights, of the 19th and early 20th centuries.

WARNING The name given to the free running of striking and chiming trains some three minutes before they actually sound. Some systems do not employ a warning and, in particular, there is no warning in a mechanism which is intended to repeat.

# LIST OF SUPPLIERS

Here are some suggestions, to which should be added advertisers from specialist magazines. Some listed here under special heads are also general suppliers of parts and tools.

*GENERAL*

Cousins, E., 335 Green Lane, Ilford, Essex IG3 9TH.

Greville, Charles, Willey Mill House, Alton Road, Farnham, Surrey GU10 5EL.

Hadfield, G. K., Blackbrook Hill House, Tickow Lane, Shepsted, Leics LE12 9EY.

Meadows & Passmore, Farningham Road, Jarus Brook, Crowborough, East Sussex TN6 2JP.

Rose, R. E., 731 Sidcup Road, Eltham, London SE9 3SA.

Southern Watch & Clock Supplies Ltd., 48/56 High Street, Orpington, Kent BR6 0JH.

Walsh & Sons Ltd., 243 Beckenham Road, Beckenham, Kent BR3 4TS.

The Watchmakers' Supply Co. Ltd., 30 The Crescent, Carterton, Oxford OX8 3SL.

J. M. Wild (Clocks), 12 Norton Green Close, Sheffield S8 8BP.

*BELLS*

Whitechapel Bell Foundry, 32/4 Whitechapel Road, London E1 1EW.

*CUTTERS*

Nathan Shestopal Ltd., Unit 2, Sapcote Trading Centre, 374 Willesden High Road, London NW10 2DH.

P. P. Thornton (Successors) Ltd., The Old Bakehouse, Upper Tyso, War CV35 0TR.

*DIALS* (to specifications)

Richards of Burton, Woodhouse Clock Works, Swadlincote Road, Woodville, Burton on Trent DE11 8DA.

Goodacre Engraving Ltd., Lodge House, Wyvern Industrial Estate, Long Eaton, Nottingham NG10 1AU.

*GLASS DOMES*

The Glass Dome Company, High Street, Leigh, nr. Tonbridge, Kent TN11 8RH.

*METALS*

Crafts for Four Seasons, 1120 Melton Road, Syston, Leicester.

Maidstone Engineering Services, 50 Hedley Street, Maidstone, Kent ME14 5AD.

J. Smith & Sons, 42–54 St John's Square, London EC1P 1ER.

K. R. Whiston Ltd., New Mills, Stockport SK12 4PT.

*OLD MOVEMENTS AND PARTS*

Allcroft, A., Bulshaw Farm, Little Hayfield, via Stockport, Cheshire.

Biddle & Mumford (Gears), 36 Clerkenwell Road, London EC1M 5PQ.

Olivers, 15 Cross Street, Hove, East Sussex BN3 1AJ.

*WOOD MOULDINGS, VENEERS*

Weaves & Waxes, 53c Church Street, Bloxham, Banbury OX15 4ET.

*WHEEL & PINION CUTTING MACHINES*

Chronos Ltd., 95 Victoria Street, St Albans, Herts.

Colin Walton Clocks, Tunbeck Cottage, Alburgh, Harleston, Norfolk IP21 0BS.

# SELECT BIBLIOGRAPHY

ALLIX, C., *Carriage Clocks*, Antique Collectors' Club, 1974.

BRITTEN, F. W., *Horological Hints and Helps*, 1929, repr. Baron 1977.

CARLE, D., *Practical Clock-Repairing*, NAG, 1952, later reprints.

GAZELEY, W. J., *Watch and Clock Repairing*, Heywood, 1965, later reprints.

JENDRITZKI, H. and MATTHEY, J. P. *Repairing Antique Pendulum Clocks*, Lausanne, 1973.

PENMAN, L., *Clock Design and Construction*, Argus, 1984.
*The Clock-Repairer's Handbook*, David & Charles, 1985.

ROBERTS, D., *The Bracket Clock*, David & Charles, 1982.

ROBINSON, T., *The Long-Case Clock*, Antique Collectors' Club, 1981.

SMITH, E., *Repairing Antique Clocks*, David & Charles, 1973, repr. 1975.
*Striking and Chiming Clocks*, David & Charles, 1985.

VERNON, J., *The Grandfather Clock Maintenance Manual*, David & Charles, 1983.

WEAVER, J. D., *Electrical and Electronic Clocks and Watches*, Newnes, 1982.

WHITEN, A. J., *Repairing Old Clocks and Watches*, NAG, 1979.

WILDING, J., *How to Repair Antique Clocks*, 4 vols, Brant Wright Associates, n.d.

WILDING, J., *A Weight Driven 8-Day Wall Clock*, Brant Wright Associates, n.d.

# INDEX